Bruce L. Golden Edward A. Wasil
Patrick T. Harker (Eds.)

The Analytic Hierarchy Process

Applications and Studies

With Contributions by

J. M. Alexander, W. D. Daniel Jr., J. G. Dolan, L. P. Fatti, B. L. Golden, R. P. Hamalainen, P. T. Harker, D. E. Levy, R. Lewis, M. J. Liberatore, E. R. MacCormac, R. J. Might, K. H. Mitchell, W. R. Partridge, J. B. Roura-Agusti, J. Ruusunen, T. L. Saaty, K. Tone, L. G. Vargas, J. G. Vlahakis, Q. Wang, E. A. Wasil, S. Yanagisawa

With 60 Figures

Springer-Verlag Berlin Heidelberg New York
London Paris Tokyo Hong Kong

Professor Bruce L. Golden
Department of Management Science and Statistics
College of Business and Management
University of Maryland
College Park, MD 20742, USA

Assistant Professor Edward A. Wasil
Kogod College of Business Administration
American University
Washington, D.C. 20016, USA

Assistant Professor Patrick T. Harker
The Wharton School
The University of Pennsylvania
Philadelphia PA 19104, USA

ISBN 978-3-642-50246-0 ISBN 978-3-642-50244-6
DOI 10.1007/978-3-642-50244-6

Softcover reprint of the hardcover 1st edition 1989

2142/7130-543210

CONTENTS

INTRODUCTION

Management science is a discipline dedicated to the development of techniques that enable decision makers to cope with the increasing complexity of our world. The early burst of excitement which was spawned by the development and successful applications of linear programming to problems in both the public and private sectors has challenged researchers to develop even more sophisticated methods to deal with the complex nature of decision making. Sophistication, however, does not always translate into more complex mathematics. Professor Thomas L. Saaty was working for the U.S. Defense Department and for the U.S. Department of State in the late 1960s and early 1970s. In these positions, Professor Saaty was exposed to some of the most complex decisions facing the world: arms control, the Middle East problem, and the development of a transport system for a Third-World country. While having made major contributions to numerous areas of mathematics and the theory of operations research, he soon realized that one did not need complex mathematics to come to grips with these decision problems, just the right mathematics! Thus, Professor Saaty set out to develop a mathematically-based technique for analyzing complex situations which was sophisticated in its simplicity. This technique became known as the Analytic Hierarchy Process (AHP) and has become very successful in helping decision makers to structure and analyze a wide range of problems.

Since Saaty's initial development of the AHP in the 1970s and the publication of his first book on the subject in 1980, numerous theoretical extensions and empirical applications have appeared in the literature. Saaty's application of the AHP to develop a plan for designing the transportation infrastructure of the Sudan, begun in 1973, is one of the earliest full-scale applications reported. In recent years, special issues of Socio-Economic Planning Sciences and Mathematical Modelling have been dedicated to the study of AHP. These journal issues and the proceedings of the first international conference dedicated solely to the AHP (held in Tianjin, China) illustrate the fact that the AHP has been accepted by the international scientific community as a very useful tool for dealing with complex decision problems. In addition, many corporations and governments are routinely using the AHP for major policy decisions.

Although there is a considerable body of literature that focuses on the use of the AHP, much of it is journal-based and therefore not easily accessible to operations research practitioners and researchers, corporate decision makers, and students. Furthermore, there are very few articles that fully describe AHP modeling and implementation issues. In fact, many applications presented in the scattered journal literature tend to be of the "arm chair" variety offering few real-world components or insights.

The purpose of this book is to provide a unified treatment of the basics of the AHP, its recent extensions, and the wide variety of potential applications to which it is suited. In particular, one of our key aims has been to assemble a collection of informative and interesting articles that focus on the application of the AHP to important, diverse, real-world decision problems.

The book is divided into three sections. In the first section, a detailed tutorial and an extensive annotated bibliography serve to introduce the methodology. The second section includes two papers which present new methodological advances in the theory of the AHP. The third section, by far the largest, is dedicated to applications and case studies; it contains twelve chapters. Papers dealing with project selection, electric utility planning, governmental decision making, medical decision making, conflict analysis, strategic planning, and others are used to illustrate how to successfully apply the AHP. Thus, this book should serve as a useful text in courses dealing with decision making as well as a valuable reference for those involved in the application of decision analysis techniques.

The AHP is being used around the world and we have sought to reflect this in the present volume. The chapter authors are mainly from the U.S., however, Europe, Asia, Canada, and South Africa are also represented. In addition, one article focuses on conflict analysis in Northern Ireland. Another discusses business strategy formulation for a financial institution in Central America.

As editors, we wish to extend a sincere thank you to each and every author. We are hopeful that they will be as proud of this volume as are the co-editors. In addition, we thank Dr. Werner A. Muller, Economics Editor for Springer-Verlag, for his encouragement and support and Irene Hagerty for her skillful help in producing the volume.

Bruce L. Golden
University of Maryland

Edward A. Wasil
American University

Patrick T. Harker
University of Pennsylvania

March 1989

THE ART AND SCIENCE OF DECISION MAKING: THE ANALYTIC HIERARCHY PROCESS

Patrick T. Harker
Decision Sciences Department
The Wharton School
University of Pennsylvania
Philadelphia, Pennsylvania 19104

ABSTRACT

This paper presents an overview of the philosophy and methodology which underlies the Analytic Hierarchy Process. After introducing the method through a series of examples, the theoretical basis of the method is described along with a summary of its mathematical underpinnings. Several recent methodological extensions are also described along with a brief description of several major and illustrative applications. The paper concludes with a summary of the progress to date in the continuing development and application of this important decision-aiding methodology.

1. SO YOU HAVE A DECISION TO MAKE!

When you are faced with a decision to make, how do you typically proceed? For most people in most circumstances, you simply decide at the particular moment based on prior experience, intuition, advice from others, etc. However, some people have a very hard time making even the most mundane decisions (spoken from experience) and, in major decisions, we all have trouble. Furthermore, even if we know with certainty what we would like to decide, we still must convince others (e.g., spouse, boss) that we know what we are doing. In this case, intuition rarely suffices; the answer "because I just want to" never worked as a teenager when we confronted our parents and surely won't work with our boss. Thus, for most decisions, we either approach the problem from a holistic point of view in which we simply choose the best, or we somehow break the decision down into components in order to (a) better understand the problem we are faced with and/or (b) communicate with someone else why a particular course of action was chosen.

For example, when confronted with the problem of buying a new car, I may know in my heart of hearts that I want the Porsche without any further analysis. Thus, a *holistic* approach in which I simply choose the preferred alternative without any analysis is very often the best method for decision-making. However, I may really want to break the decision down into the tradeoff between costs (purchase, maintenance), performance, and style to get a better understanding of my true preferences. Furthermore, such a breakdown is vital if I am ever to succeed in convincing my wife that a Porsche is really a good choice! Thus, holistic methods can often suffice, but for major decisions, one needs a more scientific/logical approach to decision-making.

The purpose of this paper is to introduce an approach to decision-making which provides the necessary logical/scientific foundations which are often required, but does not lose sight of the fact that decisions are ultimately dependent on the creative process by which the decision problem is formulated. This method, called the *Analytic Hierarchy Process* or *AHP*, was first developed by Professor Thomas L. Saaty in the 1970s and, since that time, has received wide application in a variety of areas [9]. Rather than begin this exposition of the method with a formal discussion of the underlying theory, let us consider a simple decision problem.

Let's begin with a simple estimation situation. Suppose that I am without access to an atlas and would like to estimate the relative distances of various cities with respect to their distance from Philadelphia; the cities under study are: Boston, Houston, Los Angeles, and St. Louis. How would I begin? The first question to be addressed is to decide on what type of information I can supply. If I want to compare the distances of various cities from Philadelphia, a very natural response would be to compare relative distances of pairs of cities. For example, I may estimate that Los Angeles is nine times further from Philadelphia than is Boston. Thus, I am supplying *ratio scale* judgments on the relative distance of each city pair; that is, my response to the question of how far each city is from Philadelphia is in the form of the ratio of the distances. Also, distances are not negative; thus, our responses will be limited to *positive* numbers. Furthermore, if I state that Los Angeles is nine times further from Philadelphia than is Boston, then I should agree that Boston is one-ninth as far as Los Angeles. Thus, my responses would also be *reciprocal* in the above mentioned sense. Finally, I surely must agree that the relative distance of Boston with respect to Boston is one. In summary, a very natural way in which to answer the question of comparing relative distances of cities from Philadelphia is to respond with *positive, reciprocal* judgments based on a ratio scale. A possible set of these judgments is given in Table 1.

Table 1. Judgments for the Distance to Philadelphia Example

	Boston	Los Angeles	St. Louis	Houston
Boston	1	1/9	1/3	1/4
Los Angeles	9	1	3	2
St. Louis	3	1/3	1	1/2
Houston	4	1/2	2	1

Table 2. Relative Distance Estimates

	Actual Distance (miles)	Normalized Distance	Estimated Distance
Boston	296	0.055	0.059
Los Angeles	2,706	0.503	0.513
St. Louis	868	0.161	0.160
Houston	1,508	0.280	0.269
	Sum=5,378		C.R.=0.006

In Table 1, note that we have made some "errors" when providing the judgments on the relative distances. For example, we say that Houston is 4 times further than Boston and that Los Angeles is 2 times further than Houston, which should imply that Los Angeles is 2 X 4 = 8 times further than Boston; however, we have provided a 9 for the Boston-Los Angeles judgment. In fact, this matrix of judgments has several other "errors." If no such errors existed, then we could take any one column of the above matrix and normalize it to yield the overall distances for each city. For example, taking the Boston column yields: (1/17, 9/17, 3/17, 4/17) = (0.058, 0.529, 0.160, 0.269). However, taking the Houston column provides a different estimate of the relative distances: (0.067, 0.533, 0.133, 0.267). As will be described in Section 3 and in detail in Appendix A, the AHP deals formally with these "errors" by estimating the overall weights (distances) using all of the information contained in the matrix, not just in one particular column as shown above. Using the technique described in Appendix A, the estimate of the weights has been computed and is shown in Table 2. Note that when compared to the actual distances, our simple estimation procedure has done quite well! As will be described more fully in Section 3, the number "C.R." provides a measure of how inconsistent we were in filling in the matrix. That is, the *consistency ratio C. R.* provides a way of measuring how many "errors" were created when providing the judgments; a rule-of-thumb is that if the C.R. is below 0.1, then the errors are fairly small and thus, the final estimate can be accepted. As shown in Table 2, we have been quite consistent in our judgments under this measure.

The example just presented provides an introduction to the "heart" of the AHP procedure; namely, the ability to make paired comparisons of objects with respect to a common goal or criteria (e.g., distance to Philadelphia). By understanding the above process, we have demonstrated one of the two essential components of the AHP - the *analytical* process of judgment and the creative process of constructing and analyzing a *hierarchy*. To understand the latter component, let us consider a more involved example.

Table 3. Inverse Distance to Philadelphia Data

	Inverse of Actual Distance	Normalized Inverse Distances
Boston	0.0033784	0.6073
Los Angeles	0.0003695	0.0664
St. Louis	0.0011521	0.2071
Houston	0.0006631	0.1192
	Sum=0.0055631	

Consider a recent MBA graduate from the Wharton School in Philadelphia with four job offers to consider in four different cities: Boston, Los Angeles, St. Louis, and Houston. The jobs in each of the four cities are basically equal; thus, the graduate must decide which job to take on the basis of the overall quality of life in each city. As a native Philadelphian, a major criteria in this decision is the distance of the city from the MBA's hometown. For example, consider the data in Table 2. Since the graduate has access to an atlas, estimates of the distances are replaced by the actual distance information. If larger numbers refer to "more preferred" alternatives, the distance figures are inverted and renormalized to yield the results presented in Table 3. Thus, Boston is the most preferred city and is followed in order by St. Louis, Houston, and Los Angeles. If distance were the sole criteria on which to base the job decision, then the graduate is done; Boston is the clear choice. However, other factors influence the decision:

* Cost of living
* Climate
* Educational facilities
    - Elementary and high schools
    - Colleges and universities
* "Quality of Life" factors
    - Ease of commuting to and from work
    - Arts and recreational facilities.

How does the graduate make tradeoffs between these various criteria in deciding where to move?

The first task the MBA must undertake is how to structure the decision problem. One of the easiest methods is to create a *hierarchy* of criteria, sub-criteria, and alternatives. For example, Figure 1 shows how the influential factors can be placed into the form of a hierarchy. Thus, the overall goal of choosing the best city in which to live is at the top of the hierarchy, the criteria are at the next level, followed by the subcriteria, and finally the alternatives.

Once this structuring of the problem has been finished, the next task involves the elicitation of judgments for how "good" the cities perform under each criterion. Very often, pairwise comparisons as in the above distance example are not necessary since measurements are already available with which to judge the alternatives. In the current example, Table 3 lists the data we need in order to judge the cities under the Distance criterion. However, what about criteria such as Cost of Living and Climate? One can find measures of these items in statistical abstracts and other reference books, but these measurements may have no or little relationship to what are perceived to be the best cities under these criteria. For example, the cost of living statistics are usually generated for the average consumer. Does such a person ever exist? Even if he did, does the graduate look like the average? If so, then the published cost of living statistics can be used in exactly the same way as the distance measurements. If not, then value judgments must be made on the relative "goodness" of each city with respect to that criterion. In the current example, assume that the average cost of living statistics are sufficient (see Table 4), but that the Climate rankings found in the literature are not believable. Therefore, the graduate must make value judgments of the cities with respect to climate.

In order to make the judgments on climate, some type of scale is needed. In the distance example, it was very easy to say "St. Louis is 3 times further from Philadelphia than Boston." With climate, however, what does "3 times" mean? To overcome this difficulty, let us define the verbal scale shown in Table 5; the reason for using the exact numbers in this scale will be provided in Section 3. Thus, when asked to compare Boston and Los Angeles with respect to climate, the judgment is that Los Angeles is "slightly better" than Boston. Table 6 lists the results of this questioning with respect to climate. For example, the Boston versus Los Angeles entry is 1/3 and Los Angeles-Boston entry is 3 to represent the fact that Los Angeles is slightly better; remember that the judgments are reciprocal! As one can see, the graduate feels that Los Angeles is the best city with respect to climate, and is followed in order by Boston, St. Louis, and Houston (the MBA does not like humidity!).

One can perform the same pairwise comparison procedure for all of the remaining criteria. However, note that in the case where only two subcriteria need to be compared (elementary/high-schools versus colleges; commuting versus arts and recreation), the pairwise comparison procedure is equivalent to assigning two numbers which sum to one. The results of these comparisons can be found in Tables 7-11; note that the judgments were fairly consistent throughout the procedure since the C.R. is less than 0.1 for all comparison matrices.

Figure 1. Hierarchy for the City Choice Problem

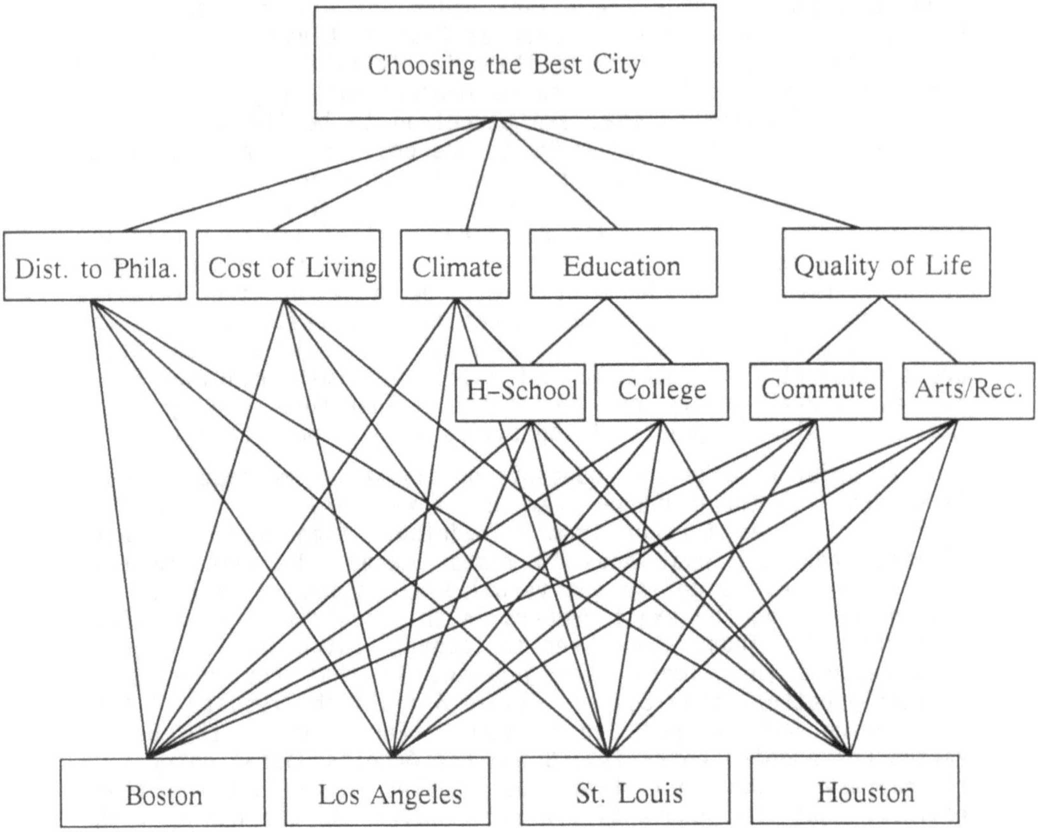

Table 4. City Rankings with Respect to Cost of Living

	Cost of Living Index (COL)	Percentage above Minimum COL	Inverse of Percentages	Normalized Inverse Percentages
Boston	335.1	1.392	0.7184	0.31836
Los Angeles	345.1	4.418	0.2263	0.10031
St. Louis	330.5	1.000	1.0000	0.44315
Houston	341.1	3.207	0.3118	0.13818
			2.0302	

Table 5. Scale of Measurement for AHP

Numerical Values	Definition
1	Equally important or preferred
3	Slightly more important or preferred
5	Strongly more important or preferred
7	Very strongly more important or preferred
9	Extremely more important or preferred
2,4,6,8	Intermediate values to reflect compromise
Reciprocals	Used to reflect dominance of the second alternative as compared with the first.

Table 6. City Comparison with Respect to Climate

	Pairwise Comparisons				
	Boston	Los Angeles	St. Louis	Houston	Relative Priority
Boston	1	1/3	2	5	0.259
Los Angeles	3	1	4	5	0.537
St. Louis	1/2	1/4	1	2	0.132
Houston	1/5	1/5	1/2	1	0.072
					C.R.=0.026

Table 7. City Comparison with Respect to Elementary and High Schools

	Pairwise Comparisons				
	Boston	Los Angeles	St. Louis	Houston	Relative Priority
Boston	1	5	1	4	0.421
Los Angeles	1/5	1	2	2	0.246
St. Louis	1	1/2	1	2	0.229
Houston	1/4	1/2	1/2	1	0.104
					C.R.=0.071

Table 8. City Comparison with Respect to Colleges and Universities

	Pairwise Comparisons				
	Boston	Los Angeles	St. Louis	Houston	Relative Priority
Boston	1	2	5	6	0.538
Los Angeles	1/2	1	2	3	0.253
St. Louis	1/5	1/2	1	2	0.130
Houston	1/6	1/3	1/2	1	0.079
					C.R.=0.006

Table 9. City Comparison with Respect to Commuting

	Pairwise Comparisons				
	Boston	Los Angeles	St. Louis	Houston	Relative Priority
Boston	1	1	1/7	1/6	0.063
Los Angeles	1	1	1/8	1/7	0.059
St. Louis	7	8	1	2	0.530
Houston	6	7	1/2	1	0.348
					C.R.=0.010

Table 10. City Comparison with Respect to Arts and Recreation

	Pairwise Comparisons				
	Boston	Los Angeles	St. Louis	Houston	Relative Priority
Boston	1	1/2	4	5	0.324
Los Angeles	2	1	5	6	0.508
St. Louis	1/4	1/5	1	2	0.103
Houston	1/5	1/6	1/2	1	0.066
					C.R.=0.015

Table 11. Comparison of Subcriteria

Subcriteria	Relative Priority
Elementary-High Schools	0.4
Colleges and Universities	0.6
Commuting	0.3
Arts and Recreation	0.7

Table 12. Comparison of Criteria with Respect to the Goal

	Pairwise Comparisons					
	Distance	Cost	Climate	Education	Quality of Life	Relative Priority
Distance	1	2	1/2	2	1/3	0.157
Cost	1/2	1	1/4	2	1/3	0.107
Climate	2	4	1	2	1	0.302
Education	1/2	1/2	1/2	1	1/4	0.088
Quality of Life	3	3	1	4	1	0.346
						C.R.=0.036

Given all of the comparisons, the next question is how to bring all of this information together to make a decision. The classical method is simply to add up the numbers under each criterion and choose that city with the highest total. In other forms, this procedure becomes the method of listing "pros" and "cons" and then choosing that alternative with the most "pros." However, the criteria are not equally weighted! Thus, we must ask how important the various criteria are to our decision. Note that this process was begun above through asking which of the subcriteria are more important. Table 12 provides the comparison matrix for the judgment of which criteria are most important with respect to the goal of choosing the best city to live in. As one can plainly see, the criteria are different.

We now have the priorities or preferences of each city under the various criteria and subcriteria as well as the relative importance of these criteria; what do we do with these numbers? The next step (which will be justified in Section 3) is to "add up" the relative priorities by weighting them with the overall priority of the given criterion. For example, Table 13 contains the "summation" of the weights of the cities with respect to the four subcriteria. This "addition" creates a composite measure of the importance or preference for each city with respect to the overall Education and Quality of Life criteria. Note that we shall also add up the consistency measures, C.R., by weighting them according to the priority of that criterion. The intuition behind this last operation is that if we are very inconsistent on a relatively unimportant criterion, it really should not matter. Finally, Table 14 contains the results of "adding" all the priorities of the cities under each criterion to yield the final composite priorities. This process of moving up the hierarchy shown in Figure 1 to yield the final weights for each alternative under the stated goal is known as *hierarchical composition or synthesis* and will be described more fully in Section 3. Without describing the mathematics, however, this procedure is fairly intuitive and should be easy to understand. Looking at the results, we see that the graduate likes the coasts! The MBA could now explore the impact of changing certain judgments on the final decision in order to get a better feel for the "robustness" of the final set of weights or priorities. This exercise in *sensitivity analysis* is a crucial step in any modeling effort and should not be ignored in a careful application of the AHP.

Readers that can follow and replicate the above example are well on their way to mastering the basics of the AHP. In the remainder of this paper, we will attempt to explain the philosophy and theory behind this method, as well as sketch some recent extensions, possible pitfalls, and interesting applications. However, if one really would like to gain an appreciation for how the method works, simply try it!

Table 13. Composite Priorities Under Education and Quality of Life

	Composite Priorities for Education			Composite Priorities for Quality of Life		
	High School (0.4)	Colleges (0.6)	Composite	Commuting (0.3)	Arts/Rec. (0.7)	Composite
Boston	0.421	0.538	0.491	0.063	0.324	0.246
Los Angeles	0.246	0.253	0.250	0.059	0.508	0.373
St. Louis	0.229	0.130	0.170	0.530	0.103	0.231
Houston	0.104	0.079	0.089	0.348	0.066	0.151
C.R.	0.071	0.006	0.032	0.010	0.015	0.014

Table 14. Composite Priorities of the Cities

	Distance (0.157)	Cost of Living (0.107)	Climate (0.302)	Education (0.088)	Quality of Life (0.346)	Composite
Boston	0.607	0.318	0.259	0.491	0.246	0.336
Los Angeles	0.066	0.100	0.537	0.250	0.373	0.335
St. Louis	0.207	0.443	0.132	0.170	0.231	0.215
Houston	0.119	0.138	0.072	0.089	0.151	0.115
C.R.	0.000	0.000	0.026	0.032	0.014	0.015

2. WHY AHP? THE ART AND SCIENCE OF DECISION MAKING

As was illustrated in the previous section, the AHP is an intuitive and relatively easy method for formulating and analyzing decisions. The city choice example illustrated the three major concepts behind the AHP: *analytic, hierarchy*, and *process*. In what follows, we would like to briefly describe the philosophy behind these three components.

Analytic. Simply put, the AHP uses numbers. Note that in holistic decision making, as described in the beginning of the previous section, no numbers are needed in order to arrive at a decision; simply choose the alternative that is most desired. However, as discussed previously, there are very good reasons why you would like to use mathematics to understand and/or describe your choice to others. In this sense of the word, all methods which seek to describe a decision are analytic since they must use mathematical/logical reasoning.

Hierarchy. The AHP structures the decision problem in levels which correspond to one's understanding of the situation: goals, criteria, subcriteria, and alternatives. The city choice example of the previous section is a relatively simple hierarchy since it consists of only four levels; examples in the literature [3,9,10,20,21,22,26,35,36,37] show the tremendous complexity which can be dealt with in a hierarchy. By breaking the problem into levels, the decision maker can focus on smaller sets of decisions; evidence from psychology suggests that humans can only compare 7 ± 2 items at a time - the so-called Miller's Law [17,31]. Thus, it is vital if we are to deal with complex situations that we use a hierarchy. In Section 5, an extension of the concept of a hierarchy will be presented which provides for even further complexity.

Process. As most know, decisions which are truly important cannot be made in a single meeting; one cannot expect the AHP to counteract this basic human tendency. People need time to think about a decision, gather new information, negotiate if it is a group decision, etc. Thus, any real decision problem involves a *process* of learning, debating, and revising one's priorities. As envisioned by Saaty, the AHP is meant to be used to aid and hopefully shorten this decision process through the insights which this method can generate; it was never and will never replace the overall decision process! The AHP points to where more information is needed, where major points of disagreement lie, etc. Also, when one goes through the structured process as in the city example, the final result may not agree with my "gut feelings"; I may really want to live in Los Angeles. At this point, a decision maker must return to the hierarchy in order to see if any true feelings have been misrepresented (the MBA likes warm weather, so maybe Climate should be given an even higher priority), or it may be that intuitive feelings will change after considering the problem in detail. This process is unavoidable and is in fact quite healthy; the AHP is meant to aid and not destroy this natural process of decision making.

Therefore, the overall philosophy of the AHP is to provide a solid, scientific method (the analytic part) to aid in the creative, artistic formulation and analysis of a decision problem. Saaty [20,21] provides a great deal of insight into both the scientific and artistic nature of this process. In what follows the reader is introduced to the scientific component of the AHP. The artistic/creative aspects are much more difficult to discuss. The papers in the remainder of this book and the studies surveyed by Golden, Wasil, and Levy in this volume provide a glimpse into the creative process of problem formulation. However, to truly learn the power of the AHP, it is necessary to go through the creative process of formulating and analyzing a problem.

3. THEORETICAL UNDERPINNINGS OF THE AHP

The AHP is based on a set of axioms which were first stated by Saaty [23] and are described in the paper by Harker and Vargas [14]. An excellent introduction to the method and its theoretical underpinnings is provided by Rozanne Saaty [18]. These basic set of assumptions provide the theoretical basis on which the method in founded. Rather than stating these assumptions in their full mathematical form, we simply paraphrase them in order to understand their meaning.

Axiom 1 Given any two alternatives (or sub-criteria) i and j out of the set of alternatives A, the decision maker is able to provide a pairwise comparison a_{ij} of these alternatives under any criterion c from the set of criteria C on a ratio scale which is reciprocal; i.e.,

$$a_{ji} = 1/a_{ij} \quad \text{for all } i,j \in A.$$

Axiom 2 When comparing any two alternatives $i,j \in A$, the decision maker never judges one to be infinitely better than another under any criterion $c \in C$; i.e., $a_{ij} \neq \infty$ for all $i,j \in A$.

Axiom 3 One can formulate the decision problem as a hierarchy.

Axiom 4 All criteria and alternatives which impact the given decision problem are represented in the hierarchy. That is, all the decision maker's intuition must be represented (or excluded) in terms of criteria and alternatives in the structure and be assigned priorities which are compatible with the intuition.

The first axiom was already discussed in Section 1. If a decision maker is able to say something is five times more important than something else, then he should agree that the reciprocal property holds. The second assumption is vital; it says that infinite preferences are not allowed. Consider a situation in which under one criterion an alternative has infinite preference. In this situation, there is really no choice under that criterion since the other alternatives will not matter at all. In this situation, one really doesn't need a decision tool; you know the answer for that criterion!

The third and fourth axioms are more subtle than the first two. Axiom 3 simply states that the problem can be formulated as in Section 1. As we shall illustrate in Section 5, not all problems fit this framework and thus, one must be very careful. If one can clearly delineate the criteria, subcriteria, alternatives, and their interrelationships, then one can most likely formulate the problem in a hierarchical fashion. The simple test to see if this axiom holds is to try to formulate the problem in a hierarchy. Axiom 4 is somewhat odd. Basically, it states that if your thinking of considering a criterion or alternative, be sure to include it. The reason for this assumption is that, as will be explained later, the AHP can exhibit *rank reversal* [4,14,24,27]. That is, the method may give one ordering of the alternatives if, for example, five alternatives are available, and a different ordering if one is dropped. The reasons why this phenomena occurs will be explained below. At this point, it is best to follow the advice of Axiom 4; include everything that matters into the decision hierarchy.

The above axioms are used to describe the two basic tasks in the AHP: formulating and solving the problem as a hierarchy (3 and 4), and eliciting judgments in the form of pairwise comparisons (1 and 2). To describe the method in some detail, let us look at each of these steps.

As illustrated by the distance example in Section 1, the elicitation of priorities for a given set of alternatives A under a given criterion $c \in C$ involves the completion of n x n matrix, where n is the number of alternatives under consideration. However, since the comparisons are assumed to be reciprocal, one needs to answer only $n(n - 1)/2$ of the comparisons to completely fill in the matrix of judgments $A = (a_{ij})$. This matrix A is *positive* and *reciprocal*. The question now before us is how to derive the overall rankings of the alternatives from the pairwise comparisons. The first and simplest method is to simply normalize one column as done in Section 1. However, when errors are permitted in eliciting the pairwise comparisons, the final answer will depend on which column is chosen for the normalization; the distance example of Section 1 demonstrates this fact.

Why allow errors from the outset? As will be discussed in the next section, all other decision-aiding methodologies require that the decision maker make no errors in providing the preference information. Thus, the ability to deal formally with judgment errors is unique to the AHP. One way to avoid errors in the AHP would be to simply ask the decision maker to compare all alternatives $i = 2,3,\ldots,n$ to alternative 1. In the distance example, we could have simply asked the decision maker to compare everything against Boston. However, why Boston and not Houston? Avoiding errors implies that we must make an a priori and ad hoc assumption on which alternative we shall treat as the base for comparison. The AHP, through the requirement of asking $n(n-1)/2$ questions, avoids this problem. Errors will always occur in judgment. We can either assume them away or deal formally with them when they occur; the latter is the philosophy underlying the AHP.

Saaty [19,20] proposed an eigenvector approach for the estimation of the weights from a matrix of pairwise comparisons A; this method is explained in detail in Appendix A. As described by Harker and Vargas [14], other methods of estimation do exist. However, evidence [8,24,27,28,30] suggests that the eigenvector approach is a theoretically and practically proven method for estimating the weights. As described in [14], the eigenvector also has an intuitive interpretation in that it is an averaging of all possible ways of thinking about a given set of alternatives. Thus, the estimation of the weights of a given set of alternatives is a well understood and easily implemented procedure.

After estimating the weights, the decision maker is also provided with a measure of the inconsistency of the given pairwise comparisons. As described in Appendix A, this *consistency ratio C.R.* provides a measure of the probability that the matrix was filled in purely at random. That is, the C.R. is a comparison of the current matrix and a purely random answering of questions. The number 0.1 which is the accepted upper limit for C.R. says that, loosely speaking, there is a 10% chance that the decision maker answered the questions in a purely random manner. With more consistent judgments, the less likely it is that the matrix was filled in at random and thus, C.R. decreases. If C.R. > 0.1, it is recommended that the decision maker revise some judgments since they are highly inconsistent; Harker [11] describes a method for choosing which judgments should be considered for revision in order to reduce inconsistency. Thus, the AHP does not require decision makers to be consistent but, rather, provides a measure of inconsistency as well as a method to reduce this measure if it is deemed to be too high.

The last question concerning the elicitation of pairwise comparisons deals with the scale of measurement. Axiom 2 requires that the pairwise comparisons a_{ij} be bounded, but it provides no guidance as to what value this bound should take. In Table 5, the upper limit is chosen to be 9. In theory, any number less than infinity can be used for the upper bound. Extensive practical experience [14,20], however, suggests that 9 is a good upper bound to use. Thus, the scale presented in Table 5 is the suggested scale unless the decision maker feels more comfortable with another or has some prior knowledge as to which scale is best in a given decision-making context.

After generating a set of weights w_a^c for each alternative $a \in A$ under a criterion $c \in C$, the *principle of hierarchic composition* provides a way of computing the overall priority of the alternatives by summing the priority under criterion c times the priority of criterion c (i.e., $w_a^c \times v_c$) or

$$w_a = \sum_{c \in C} v_c w_a^c .$$

Thus, a linear, additive function is used to represent the composite priorities of the alternatives [4]. As discussed in

[14], this principle arises from a more general approach for synthesizing judgments which will be outlined in Section 5. However, this principle is very intuitive and straightforward and thus, should be easily understood by the decision maker. For further information on this principle, the reader is referred to Saaty [20,21].

While the principle of hierarchic composition is intuitive, one must be very careful in its use. Take, for example, the following situation which is adapted from [5]. Suppose that you are deciding between the purchase of a Chevrolet, a Honda, and a Porsche under the following criteria: Cost, Style, and Performance; the results of these comparisons are listed in Table 15.

Now suppose that another Honda comes onto the market which is identical to the previous Honda except for the color and assume that color is not an important factor. The problem now has four alternatives; the results of the augmented example are shown in Table 16. Thus, rank has reversed between the Hondas and the Porsche. Why did this occur? The problem in the current example, which will be a problem in any ratio scale method, is that we have added nothing to distinguish the two Hondas. Thus, this *irrelevant alternative* has altered the overall ordering of the alternatives. The second Honda "spreads" the overall priority of this choice between two rather than one alternative. Thus, if an alternative is a copy of another or nearly so with respect to the given criteria, either the set of criteria should be revised, or this alternative should be deleted from the choice set. That is, the set of alternatives should form a *basis* on which all other alternatives can be measured. Only those alternatives which are truly unique should be kept in the decision hierarchy. In the current example, the second Honda can be deleted and then one can measure its overall priority by the first. Again, if something is not a unique alternative, don't consider it!

In summary, the AHP has a simple yet elegant theoretical foundation. However, be careful not to oversimplify. As the above example and the example to be discussed in Section 5 illustrate, ratio scales are tricky and thus, one must be careful to heed the warnings imbedded in the axiomatic foundations of the method.

Table 15. Comparison of Chevrolet, Honda, and Porsche

Cost Comparisons	Chevrolet	Honda	Porsche	Priority
Chevrolet	1	1/9	1	1/11
Honda	9	1	9	9/11
Porsche	1	1/9	1	1/11
				C.R.=0

Style Comparisons	Chevrolet	Honda	Porsche	Priority
Chevrolet	1	1	1/9	1/11
Honda	1	1	1/9	1/11
Porsche	9	9	1	9/11
				C.R.=0

Performance Comparisons	Chevrolet	Honda	Porsche	Priority
Chevrolet	1	1/9	1/8	1/18
Honda	9	1	9/8	9/18
Porsche	8	8/9	1	8/11
				C.R.=0

Criteria Comparisons	Cost	Style	Performance	Priority
Cost	1	1	1	1/3
Style	1	1	1	1/3
Performance	1	1	1	1/3
				C.R.=0

Composite Priorities				
	Cost (1/3)	Style (1/3)	Performance (1/3)	Priority
Chevrolet	1/11	1/11	1/11	0.08
Honda	9/11	1/11	9/11	0.47
Porsche	1/11	9/11	8/11	0.45

Table 16. Augmented Comparison with Two Hondas

Cost Comparisons	Chevrolet	Honda1	Honda2	Porsche	Priority
Chevrolet	1	1/9	1/9	1	1/20
Honda1	9	1	1	9	9/20
Honda2	9	1	1	9	9/20
Porsche	1	1/9	1/9	1	1/20
					C.R.=0

Style Comparisons	Chevrolet	Honda1	Honda2	Porsche	Priority
Chevrolet	1	1	1	1/9	1/12
Honda1	1	1	1	1/9	1/12
Honda2	1	1	1	1/9	1/12
Porsche	9	9	9	1	9/12
					C.R.=0

Performance Comparisons	Chevrolet	Honda1	Honda2	Porsche	Priority
Chevrolet	1	1/9	1/9	1/8	1/27
Honda1	9	1	1	9/8	9/27
Honda2	9	1	1	9/8	9/27
Porsche	8	8/9	8/9	1	8/27
					C.R.=0

Composite Priorities				
	Cost (1/3)	Style (1/3)	Performance (1/3)	Priority
Chevrolet	1/20	1/12	1/27	0.06
Honda1	9/20	1/12	9/27	0.29
Honda2	9/20	1/12	9/27	0.29
Porsche	1/20	9/12	8/27	0.37

4. COMPARISON WITH OTHER DECISION MAKING METHODOLOGIES

As discussed in Section 1, the simplest and most efficient decision making methodology is the holistic approach. If you know what you want, choose it! However, as stated in Section 1, you may not really know what you want and/or may need to convince someone else that you are making the correct choice. In these cases, the AHP philosophy of decomposing the problem into manageable subcomponents becomes vital, as does its analytic nature in using mathematical reasoning to uncover and explain one's decision.

The second most popular method of decision making is also analytical and does decompose the problem into subcomponents: the method of listing *pros and cons*. Most decision makers have used this method at some point: take a piece of paper, list the positive aspects of an alternative on one side, the negative on the other, and choose the alternative with the most "pros." As discussed in Section 1, the major problem with this technique is that it implicitly assumes that all the "pros" and all of the "cons" are equally important! For most problems, this assumption is simply not true and thus, a method like the AHP which permits the unequal weighting of criteria must be used. Decision matrix techniques [15] can be considered as extensions of the simple "pros and cons" approach and thus, are subject to the same conceptual problem.

Another very popular method for group decision making is the *Delphi technique* in which a group of decision makers are asked either through a questionnaire or through a one-on-one interview to state their preferences on a set of alternatives and these results are then statistically analyzed to yield the final outcome. As described by Saaty [20], the AHP differs from this method in three important ways.

* **Individual versus Group**. The Delphi method treats individuals separately; the AHP is used in group decision situations by having the entire group together in a single session in order to make the decision. The presence in a group allows for debate and learning on the part of the group members.

* **Serial versus Dynamic**. The Delphi technique starts with the questionnaire, analyzes the results, and then states the final decision in a step-by-step manner. The AHP, by treating the group as a whole, allows for a dynamic discussion, revision of judgments, addition or deletion of alternatives, etc. Thus, the AHP is a natural vehicle for structuring group debate; this is in fact one of its most valuable assets in practice.

* **Questionnaire versus Hierarchy**. The AHP allows the group to define, revise, and analyze the problem through the construction of the decision hierarchy. The Delphi method assumes that the analyst will structure the problem. In most cases, groups feel more comfortable with structuring their

own decision problem instead of leaving this task to a so-called expert analyst!

The final decision method which competes with the AHP is *multi-attribute utility theory* or *MAUT* [16]. Vargas [34] has shown how these two methods are related. Basically, MAUT requires the decision maker to answer questions dealing with probabilities, while the AHP uses ratio scale estimation. In some contexts, the probabilistic questions are very natural (e.g., in assessing risky assets, statistical quantities, etc.), but in others they are very unnatural. For example, in the distance estimation of Section 1, the AHP-type of questions seem very natural whereas a von Neumann-Morgenstern utility function estimation, which is the essential component of MAUT, would seem very forced. Also, MAUT implicitly assumes that the decision maker will never be inconsistent. As discussed in Section 3, this is a very strong assumption which can also make the elicitation of preferences highly biased. The AHP is the only decision-aiding methodology which deals formally with inconsistency.

In summary, other decision-making tools exist which can be useful in their proper contexts. The reader should be careful to learn about each of these methods and understand where they would fit and where they would not. While being more formal than simple pro-con choices and not dealing directly with probability/risk as in MAUT, the AHP does provide an efficient and effective tool for decision making in a wide range of problem contexts as the papers in the remainder of this book will attest.

5. EXTENSIONS OF THE BASIC METHODOLOGY

In this section, five useful extensions of the basic AHP methodology will be summarized. In all cases, the cited references contain the details for implementing these results.

Suppose that a group of decision makers is trying to elicit their judgments at a particular level of the hierarchy when a disagreement arises as to the value of a particular judgment a_{ij}. What should be done? The first attempt should be to debate why this disagreement has occurred and, if possible, to reach a consensus. However, this may not be possible. If the disagreements are large, then you may consider breaking the problem into parallel hierarchies, one for each competing faction. In this way, one can analyze what each faction would decide and see if these final decisions are very different. People may often disagree violently on certain judgments, but it ends up that these judgments have little or no impact on the final decisions. Finally, if the disagreements are small, you may consider taking the average of the group's judgments in order to save time. However, do not take the arithmetic mean! As shown in [1,2], the proper averaging method is the geometric mean. If $a_{ij}^1, a_{ij}^2, \ldots, a_{ij}^l$ represent the different judgments of the l members of the group, the composite judgment is given by:

$$a_{ij}^{avg} = [a_{ij}^{1} \times a_{ij}^{2} \times \ldots \times a_{ij}^{l}]^{(1/l)}.$$

Suppose that a group or individual cannot state a single number for a particular judgment a_{ij}. In this case, Saaty and Vargas [29,32] have developed techniques so that a range of values for a particular judgment can be stated (e.g., a_{ij} is between 2 and 3). A distribution for the resulting weights w can then be derived. While useful, this problem requires substantial research in order to create simple and efficient methods for use in routine decision making.

Harker [12,13] describes a set of techniques to reduce the number of pairwise comparisons that the decision maker must make during the analysis of a large hierarchy. For example, consider a hierarchy with 4 levels and 6 alternatives on each level. In this case, the decision maker must answer (4 X 6 X 5)/2 = 60 questions! The techniques described in Appendix C are meant to greatly reduce this number.

DeTurck [6] presents an interesting twist on the AHP methodology. The standard view of how the AHP is used is to state judgments and then estimate the weights. However, one often would like to reverse this process in practice. For example, suppose the distance judgments of Section 1 are made and the relative distance estimates shown in Table 2 are derived. However, the estimated distance for Houston is too low. What are the set of judgments closest to those given in Table 1 which would make the Houston value 0.280? DeTurck [6] provides a partial answer to this question. This technique will be very useful in situations where the AHP is used to justify a given decision in that the decision maker typically knows the desired final outcome, but would like the comparison matrix in order to illustrate the logic behind this decision.

The final extension involves the relaxation of Axiom 3 in order to allow for non-hierarchical structuring of the decision problem. This technique, known as the *system with feedback* or the *supermatrix technique*, is summarized in Appendix B. In what follows, let us consider a simple example from [7] which illustrates the trouble when a non-hierarchical problem is treated in a hierarchical fashion; see [14] for further details on this example.

Consider the problem of choosing how much money to invest in each of four alternatives A_1,A_2,A_3,A_4. The criteria for the problem are the returns in each of four years C_1,C_2,C_3,C_4. Assuming no discounting, Table 17 shows that A_2 is the best choice.

Table 17. Data for Investment Example

Alternatives	Criteria C_1	C_2	C_3	C_4	Total Return	Relative Return	Rank
A_1	1	9	1	3	14	0.2000	4
A_2	9	1	9	1	20	0.2857	1
A_3	8	1	4	5	18	0.2571	2
A_4	4	1	8	5	18	0.2571	2

Table 18. Hierarchic Composition for Investment Example

Alternatives	Weights Under Each Criteria C_1	C_2	C_3	C_4	Composite Weight	Rank
A_1	1/22	9/12	1/22	3/14	0.264	1
A_2	9/22	1/12	9/22	1/14	0.243	4
A_3	8/22	1/12	4/22	5/14	0.246	2
A_4	4/22	1/12	8/22	5/14	0.246	2

If the principle of hierarchic composition is applied directly as defined in Section 3, then each of the criteria (years) will be equally weighted since there is no discounting and the resulting relative weights of each alternative are given in Table 18. Note that rank has reversed between A_1 and A_2 and, thus, one would be tempted to conclude that the AHP is flawed. What went wrong?

The problem with the above example is that the principle of hierarchic composition does not apply in this case. If it did, then we would be able to first weight the criteria (years), and then the alternatives under each year. In order to do this, the weight or priority of the criteria must be independent of the alternative investments; in this case they are not. Even though there is no discounting, the total return under each year is different! To see this fact, consider for the moment "flipping" the problem in terms of treating the investment alternatives as criteria and the years as the alternatives. For example, answer the following question: "For alternative 1, in which year did it perform best?" If the answer for each alternative and each year is the same (i.e., year 1 is always twice as good as year 2 for all alternatives, etc.), then the priorities of the years can be generated independently from the alternatives. In the current situation, the answers will depend on the different alternatives as shown in Table 19.

In this situation, which are the criteria and which are the alternatives? In fact, it is impossible to say because we have a *system with feedback* as defined in [20] rather than a hierarchy. That is, the years depend on the alternatives in order to define their priorities and vice versa. Using the supermatrix technique outlined in Appendix B, we can define two clusters of objects (years and investments), and construct the supermatrix shown in Table 20. As described in Appendix B, the overall priorities of each element are obtained by raising the matrix to odd powers to yield the matrix given in Table 21.

Table 19. Comparison of Years with Respect to Alternatives

Years (Criteria)	Weights Under Each Alternative A_1	A_2	A_3	A_4
C_1	1/14	9/20	1/18	3/18
C_2	9/14	1/20	9/18	1/18
C_3	8/14	1/20	4/18	5/18
C_4	4/14	1/20	8/18	5/18

Table 20. Supermatrix W for the Investment Example

	C_1	C_2	C_3	C_4	A_1	A_2	A_3	A_4
C_1	0	0	0	0	1/14	9/20	8/18	4/18
C_2	0	0	0	0	9/14	1/20	1/18	4/18
C_3	0	0	0	0	1/14	9/20	4/18	8/18
C_4	0	0	0	0	3/14	3/20	5/18	5/18
A_1	1/22	9/12	1/22	3/14	0	0	0	0
A_2	9/22	1/12	9/22	1/14	0	0	0	0
A_3	8/22	1/12	4/22	5/14	0	0	0	0
A_4	4/22	1/12	8/22	5/14	0	0	0	0

Table 21. $\lim_{k\to\infty} W^{2k+1}$ for the Investment Example

	C_1	C_2	C_3	C_4	A_1	A_2	A_3	A_4
C_1	0	0	0	0	0.3143	0.3143	0.3143	0.3143
C_2	0	0	0	0	0.1714	0.1714	0.1714	0.1714
C_3	0	0	0	0	0.3143	0.3143	0.3143	0.3143
C_4	0	0	0	0	0.2000	0.2000	0.2000	0.2000
A_1	0.2000	0.2000	0.2000	0.2000	0	0	0	0
A_2	0.2857	0.2857	0.2857	0.2857	0	0	0	0
A_3	0.2571	0.2571	0.2571	0.2571	0	0	0	0
A_4	0.2571	0.2571	0.2571	0.2571	0	0	0	0

As one can see, the overall priorities of the years are given by the vector

$$(0.3143,\ 0.1714,\ 0.3143,\ 0.2000)$$

which coincides with the normalization of the total investment per year with the total investment for four years, or

$$(22/70,\ 12/70,\ 22/70,\ 14/70).$$

Also, the final priorities for the investments are identical to those listed in Table 17. In applying the AHP, a decision maker must be careful: not all decision problems can be modeled as hierarchies! The supermatrix technique allows one to break out of a hierarchical structure when needed. A simple test for when this supermatrix technique is necessary is to ask the "reverse" question of comparing elements of the next highest level in the hierarchy with respect to an element of the current level. If the question makes sense and the answers vary across elements of the current level (alternatives), then you have a system with feedback and must use the supermatrix technique.

6. APPLICATIONS

Numerous applications of the AHP have been made since its development in the mid 1970s; the papers in this volume as well as those reviewed in [9] contain a wealth of expertise in the application of the AHP to a wide variety of decision problems. Rather than reiterate these applications, we would simply like to relate three anecdotes which point to the need for AHP in many decision contexts.

In the analysis of nuclear versus non-nuclear energy in Finland [10], a debate arose in the Parliament as to whether to construct a new nuclear power plant or not. A major point of rhetoric in this debate, as is usual in Finland, centered on the Soviet influence since the nuclear plant would have been purchased from the U.S.S.R. After applying the AHP *within the Parliament* (which is a story in and of itself), most members came to realize that their rhetoric was simply that: rhetoric. Few really cared about the Soviet issue relative to other issues on the table and, thus, the debate was able to proceed without this hindrance. **The moral of the story**: The AHP can sometimes be very effective in "cutting the rhetoric" out of the debates which can often arise in group settings. In many situations, this benefit is sufficient to warrant the time expense of using the AHP.

The second incident has to do with a group application of the AHP at a major government organization. After several days of using the AHP to structure the debate surrounding a particularly touchy issue, no consensus was reached. The decision analysts (author included) felt that this application was a total disaster. However, the "boss" was very pleased since he never expected consensus; he simply wanted to observe the structured debate in order to assess which people he could trust for a more thorough application. **The moral of the story**: You needn't make a decision to have a decision-aiding methodology such as the AHP prove to be useful and insightful.

The last anecdote concerns a major international company which was facing a very risky decision. After hiring a consultant to analyze this decision with MAUT, the analysis was presented to the board. After this presentation and subsequent discussions, the decision makers were unable to understand how the numbers were obtained. The consultant told this tale of how complex the theory was and how they probably would never fully understand. The decision makers immediately found someone who had a method which they could understand - the AHP. After understanding the basics, the decision makers were then able to go ahead and apply the AHP to analyze their decision. **The moral of the story**: If the CEO doesn't understand your analysis, he is most likely not stupid; either your skills as a consultant or your underlying methodology is at fault! The AHP is a methodology which was designed to be natural and understandable. Hopefully such situations can be avoided.

7. WHERE DO WE GO FROM HERE?

The AHP is a simple yet elegant method for structuring and analyzing decisions. This paper has attempted to sketch the basics of this method as well as providing overviews of current research directions and the potholes which one must attempt to avoid on the road to successful applications. Future research should be devoted to the following areas.

* Better methods for estimating weights and reducing the burden involved in eliciting preferences. The scientific/analytic component of the *process* should not be primary in the decision maker's mind; rather, the creative/artistic process of structuring the problem and generating the alternatives should be the main component of any AHP application.

* Further study on the uses and limitations of the supermatrix technique in practice.

* A better understanding of the relationships between MAUT and the AHP and, possibly, some synthesis of these two competing schools of decision analysis.

* Improved software support for the method. Currently, *Expert Choice* is the only major microcomputer implementation of the AHP on the market in the United States. While this software is extremely useful, it does not contain any of the extensions listed in Section 5.

* The innovative application of the AHP in new and exciting areas in order to better understand the method's applicability and to provide new and interesting methodological questions.

Now that you have the basics, take the method for a spin. Happy decision making!

8. ACKNOWLEDGMENTS

This research was supported by the National Science Foundation under the Presidential Young Investigator Award ECE-8552773. The comments of Saul Gass, Bruce Golden, Tom Saaty, and Ed Wasil have greatly improved the exposition; their help is warmly acknowledged.

9. REFERENCES

1. J. Aczel and C. Alsina, "Synthesizing Judgements: A Functional Equations Approach," **Mathematical Modelling, 9**, 311-320 (1987).

2. J. Aczel and T.L. Saaty, "Procedures for Synthesizing Ratio Judgements," **Journal of Mathematical Psychology, 27**, 93-102 (1983).

3. J.F. Bard, "A Multiobjective Methodology for Selecting Subsystem Automation Options," **Management Science, 32**, 1628-1641 (1986).

4. V. Belton, "A Comparison of the Analytic Hierarchy Process and a Simple Multi-attribute Value Function," **European Journal of Operational Research, 26**, 7-21 (1986).

5. V. Belton and T. Gear, "On a Shortcoming of Saaty's Method of Analytic Hierarchies," **Omega, 11**, 228-230 (1984).

6. D.M. DeTurck, "The Approach to Consistency in the Analytic Hierarchy Process," **Mathematical Modelling, 9**, 345-352 (1987).

7. J.S. Dyer and R.E. Wendell, "A Critique of the Analytic Hierarchy Process," Working Paper 84/85-4-24, Department of Management, The University of Texas, Austin, Texas (1985).

8. J. Fichtner, "On Deriving Priority Vectors from Matrices of Pairwise Comparisons," **Socio-Economic Planning Sciences, 20**, 341-345 (1986).

9. B.L. Golden, E.A. Wasil and D.E. Levy, "Applications of the Analytic Hierarchy Process: A Categorized, Annotated Bibliography," in this volume.

10. R.P. Hamalainen and T.O. Seppalainen, "The Analytic Network Process in Energy Policy Planning," **Socio-Economic Planning Sciences, 20**, 399-405 (1986).

11. P.T. Harker, "Derivatives of the Perron Root of a Positive Reciprocal Matrix: With Application to the Analytic Hierarchy Process," **Applied Mathematics and Computation, 22**, 217-232 (1987).

12. P.T. Harker, "Incomplete Pairwise Comparisons in the Analytic Hierarchy Process," **Mathematical Modelling, 9**, 837-848 (1987).

13. P.T. Harker, "Alternative Modes of Questioning in the Analytic Hierarchy Process," **Mathematical Modelling, 9**, 353-360 (1987).

14. P.T. Harker and L.G. Vargas, "Theory of Ratio Scale Estimation: Saaty's Analytic Hierarchy Process," **Management Science, 33**, 1383-1403 (1987).

15. P. Hill et al., **Making Decisions**, Addison-Wesley, Reading, Massachusetts (1979).

16. R.L. Keeney and H. Raiffa, **Decisions with Multiple Objectives: Preference and Value Tradeoffs**, John Wiley, New York, New York (1976).

17. G.A. Miller, "The Magical Number, Seven, Plus or Minus Two: Some Limitations on our Capacity for Processing Information," **Psychology Review, 63**, 81-97 (1956).

18. R.W. Saaty, "The Analytic Hierarchy Process - What It Is and How It Is Used," **Mathematical Modelling, 9**, 161-176 (1987).

19. T.L. Saaty, "A Scaling Method for Priorities in Hierarchical Structures," **Journal of Mathematical Psychology, 15**, 234-281 (1977).

20. T.L. Saaty, **The Analytic Hierarchy Process**, McGraw-Hill, New York, New York (1980).

21. T.L. Saaty, **Decision Making for Leaders**, Lifetime Learning Publications, Belmont, California (1982).

22. T. L. Saaty, "Absolute and Relative Measurement with the AHP: The Most Livable Cities in the United States," **Socio-Economic Planning Sciences, 20**, 327-331 (1986).

23. T.L. Saaty, "Axiomatic Foundation of the Analytic Hierarchy Process," **Management Science, 32**, 841-855 (1986).

24. T.L. Saaty, "Rank Generation, Preservation and Reversal in the Analytic Hierarchy Process," **Decision Sciences, 18**, 157-177 (1987).

25. T.L. Saaty and M. Takizawa, "Dependence and Independence: From Linear Hierarchies to Non-linear Networks," **European Journal of Operational Research, 26**, 229-237 (1986).

26. T.L. Saaty and L.G. Vargas, **The Logic of Priorities**, Kluwer Nijhoff, Boston, Massachusetts (1981).

27. T.L. Saaty and L.G. Vargas, "Inconsistency and Rank Preservation," **Journal of Mathematical Psychology, 28**, 205-214 (1984).

28. T.L. Saaty and L.G. Vargas, "Comparison of Eigenvalue, Logarithmic Least Square and Least Square Methods in Estimating Ratios," **Mathematical Modelling, 5**, 309-324 (1984).

29. T.L. Saaty and L.G. Vargas, "Uncertainty and Rank Order in the Analytic Hierarchy Process," **European Journal of Operational Research, 32**, 107-117 (1987).

30. P.J.H. Shoemaker and C.C. Waid, "An Experimental Comparison of Different Approaches to Determining Weights in Additive Utility Models," **Management Science, 28**, 182-196 (1982).

31. S.S. Stevens, "On the Psychophysical Laws," **Psychological Review, 64**, 153-181 (1957).

32. L.G. Vargas, "Reciprocal Matrices with Random Coefficients," **Mathematical Modelling, 3**, 69-81 (1982).

33. L.G. Vargas, "Analysis of Sensitivity of Reciprocal Matrices," **Applied Mathematics and Computation, 12**, 201-220 (1983).

34. L.G. Vargas, "Priority Theory and Utility Theory," **Mathematical Modelling, 9**, 381-385 (1987).

35. Y. Wind, "An Analytic Hierarchy Process Based Approach to the Design and Evaluation of a Marketing Driven Business and Corporate Strategy," **Mathematical Modelling, 9**, 285-291 (1987).

36. Y. Wind and T.L. Saaty, "Marketing Applications of the Analytic Hierarchy Process," **Management Science, 26**, 641-658 (1980).

37. F. Zahedi, "The Analytic Hierarchy Process - A Survey of the Method and Its Applications," **Interfaces, 16**, 96-108 (1986).

10. APPENDIX A: MATHEMATICAL FOUNDATIONS OF THE AHP

In this appendix, the basic mathematical concepts used in the AHP will be summarized; for a more thorough treatment of these issues, the reader is referred to [19,20,21].

The first major task in the AHP involves the estimation of the weights of a set of objects (criteria or alternatives) from a matrix of pairwise comparisons $A = (a_{ij})$ which is positive and reciprocal. Thus, given the matrix

$$A = \begin{bmatrix} a_{11} & a_{12} & \cdots & a_{1n} \\ a_{21} & a_{22} & \cdots & a_{2n} \\ \vdots & \vdots & \cdots & \vdots \\ a_{n1} & a_{n2} & \cdots & a_{nn} \end{bmatrix}$$

where

$$a_{ij} = 1/a_{ji} \quad \text{for all } i,j = 1,2,\ldots,n,$$

we would like to compute a vector of weights or priorities $w = (w_1, w_2, \ldots, w_n)$. Note that by using ratio scales, the weights we estimate are only unique up to multiplication by a positive constant; i.e., w is equivalent to cw where $c > 0$. Thus, we typically will normalize w so that it sums to 1 or 100 for convenience.

If the judgments were perfectly consistent, i.e.,

$$a_{ik}a_{kj} = a_{ij} \quad \text{for all } i,j,k = 1,2,\ldots,n,$$

then the entries of the matrix A would contain no errors and could be expressed as

$$a_{ij} = w_i/w_j.$$

To see this last result, note that

$$a_{ik}a_{kj} = w_iw_k/w_kw_j = w_i/w_j = a_{ij} \quad \text{for all } i,j,k, = 1,2,\ldots,n.$$

In this case, simply normalize any column j of A to yield the final weights:

$$w_i = a_{ij}/(\Sigma_{k=1}^{n} a_{kj}) \text{ for all } i = 1,2,\ldots,n.$$

However, errors in judgment are typically made and, therefore, the final result using the column normalization would depend on which column was chosen. Two competing methods exist for estimating the weights when errors in judgment exist [8,27,24]: logarithmic least squares (LLS) and Saaty's [19] eigenvector method. LLS estimates the weights w as those which minimize the following objective:

$$\sum_{i=1}^{n} \sum_{j=1}^{n} (\ln a_{ij} - \ln w_i + \ln w_j)^2.$$

Saaty's method computes **w** as the principal right eigenvector (or Perron right vector) of the matrix A:

$$A\mathbf{w} = \lambda_{max}\mathbf{w},$$

where λ_{max} is the maximum eigenvalue (Perron root) of the matrix, or

$$w_i = \frac{\Sigma_{j=1}^{n} a_{ij}w_j}{\lambda_{max}} \quad \text{for all } i = 1,2,\ldots,n.$$

As discussed in Harker and Vargas [14] and by Fichtner [8], both methods have their advantages. However, as shown in [14], the eigenvector method has the interpretation of being a simple averaging process by which the final weights **w** are taken to be the average of all possible ways of comparing the alternatives. Thus, the eigenvector is a "natural" method for computing the weights. Furthermore, some theoretical evidence [24,27] suggests that this method is the best at uncovering the true rank-order of a set of alternatives.

The eigenvector method also yields a natural measure for inconsistency. As shown by Saaty [19,20], λ_{max} is always greater than or equal to n for positive, reciprocal matrices, and is equal to n if and only if A is a consistent matrix. Thus, $\lambda_{max} - n$ provides a useful measure of the degree of inconsistency. Normalizing this measure by the size of the matrix, Saaty defines the *consistency index (C.I.)* as:

$$C.I. = \frac{\lambda_{max} - n}{n - 1}.$$

For each size of matrix n, random matrices were generated and their mean C.I. value, called the *random index (R.I.)*, was computed; these values are illustrated in Table 22. Using these values, the *consistency ratio (C.R.)* is defined as the ratio of the C.I. to the R.I.; thus, C.R. is a measure of how a given matrix compares to a purely random matrix in terms of their C.I.'s. Therefore,

$$C.R. = C.I./R.I.$$

A value of the C.R. ≤ 0.1 is typically considered acceptable; larger values require the decision maker to reduce the inconsistencies by revising judgments [11].

Table 22. Random Inconsistency Index (R.I.)

n	1	2	3	4	5	6	7	8	9	10	11	12	13	14	15
R.I.	0.00	0.00	0.58	0.90	1.12	1.24	1.32	1.41	1.45	1.49	1.51	1.48	1.56	1.57	1.59

The computation of the principal right eigenvector is accomplished by raising the matrix A to increasing powers k and then normalizing the resulting system:

$$w = \lim_{k \to \infty} A^k e / e^T A^k e$$

where e = (1,1,...,1). For example, consider the Distance to Philadelphia example in Section 1 where

$$A^1 = A = \begin{bmatrix} 1 & 1/9 & 1/3 & 1/4 \\ 9 & 1 & 3 & 2 \\ 3 & 1/3 & 1 & 1/2 \\ 4 & 1/2 & 2 & 1 \end{bmatrix}.$$

If one normalizes each column, the following estimates of the weights are obtained:

$$\begin{bmatrix} 0.0588 & 0.0571 & 0.0526 & 0.0667 \\ 0.5294 & 0.5143 & 0.4737 & 0.5333 \\ 0.1765 & 0.1714 & 0.1579 & 0.1333 \\ 0.2353 & 0.2571 & 0.3158 & 0.2667 \end{bmatrix}.$$

Note that each column yields a different estimate of the weights. Applying the first iteration of the algorithm stated above yields:

$$w^1 = A^1 e / e^T A^1 e \quad = (0.05837,\ 0.51675,\ 0.16651,\ 0.25837).$$

Raising A to the second power $A^2 = A \times A$ yields

$$A^2 = \begin{bmatrix} 4 & 0.4583 & 1.5 & 0.8889 \\ 35 & 4 & 13 & 7.75 \\ 11 & 1.25 & 4 & 2.4167 \\ 18.5 & 2.1111 & 6.8333 & 4 \end{bmatrix},$$

which in turn creates the second estimate of the weights:

$$w^2 = A^2 e / e^T A^2 e = (0.05867,\ 0.51196,\ 0.15994,\ 0.26943).$$

Continuing this process, we have

$$w^3 = A^3 e / e^T A^3 e = (0.05882,\ 0.51259,\ 0.15958,\ 0.26890)$$

$$w^4 = A^4 e / e^T A^4 e = (0.05882,\ 0.51261,\ 0.15971,\ 0.26886)$$

$$w^5 = A^5 e / e^T A^5 e = (0.05882,\ 0.51261,\ 0.15971,\ 0.26886).$$

Thus, the process has converged in five iterations. The intuition behind this approach and its interpretation as an averaging process can be found in [14].

Once we have computed the weights as $w = w^5$, the consistency measure can be computed as follows:

$$\lambda_{max} = (\sum_{j=1}^{4} a_{1j}w_j)/w_1 = 4.01636$$

$$C.I. = (4.01636 - 4)/3 = 0.0054667$$

$$C.R. = 0.0054667/0.90 = 0.006.$$

Thus, this matrix is very consistent. As a general rule, the more inconsistent the matrix, the greater the errors in the matrix and, thus, the longer the computational procedure will take to converge.

Vargas [33] and Harker [11] provide details on computing the sensitivity of the final weights to changes in the matrix entries; Harker also shows how these results can be used to aid the decision maker in choosing which judgments to alter if the matrix is highly inconsistent.

The second major task in the analysis of a hierarchy is the synthesis of the judgments throughout the hierarchy in order to compute the overall priorities of the alternatives with respect to the goal. Saaty [20] describes the *principle of hierarchic composition* in detail and Harker and Vargas [14] show how this principle is a special case of the supermatrix technique described in Appendix B. This principle simply states that the weights are created by summing the priority of each element according to a given criteria by the weight of that criteria. The interested reader is referred to [18,20,21] for further details of this procedure.

11. APPENDIX B: THE SUPERMATRIX TECHNIQUE

The supermatrix technique which was employed in Section 5 was initially developed by Saaty [20] and has been used in a variety of contexts [10,25]. The basics of the method are as follows. Let us consider breaking the problem under study into N clusters $C_1, C_2, \ldots, C_N$. For instance, the example of Section 5 breaks the problem into a criteria set C_1 and an alternative set C_2. For each cluster i, let n_i denote the number of elements it contains, and define w_{ik}^{jl} to be the weight of element k in cluster i when compared according to the lth element of cluster j. The matrix of comparisons of the elements of cluster i with respect to the elements of j are given by:

$$W_{ij} = \begin{bmatrix} w_{i1}^{j1} & w_{i1}^{j2} & \cdots & w_{i1}^{jn_j} \\ w_{i2}^{j1} & w_{i2}^{j2} & \cdots & w_{i2}^{jn_j} \\ \vdots & \vdots & \cdots & \vdots \\ w_{in}^{j1} & w_{in}^{j2} & \cdots & w_{in}^{jn_j} \end{bmatrix} .$$

The overall matrix which contains all of the information on the comparison of all clusters with respect to every other criteria is called the *supermatrix* and is given by:

$$W = \begin{bmatrix} W_{11} & W_{12} & \cdots & W_{1N} \\ W_{21} & W_{22} & \cdots & W_{2N} \\ \vdots & \vdots & \cdots & \vdots \\ W_{N1} & W_{N2} & \cdots & W_{NN} \end{bmatrix} .$$

The overall priorities of each element of each cluster is given by the solution to:

$$\lim_{k \to \infty} W^{2k+1} .$$

Saaty [20] describes the theory behind the use of the supermatrix which is related to the theory of stochastic matrices and Markov processes. The reader is referred to the example of Section 5 to understand the mechanics of the supermatrix method and to Saaty [20] for further details on the mathematics underlying this technique.

12. APPENDIX C: ALTERNATIVE MODES OF QUESTIONING IN THE AHP

Harker [12,13] has developed several techniques to make the mechanics of the AHP easier. In particular, these methods reduce the amount of work needed to compare elements at each level of the hierarchy and allow for nonlinear responses. In this appendix, these techniques will be summarized.

One major drawback of the AHP is that at each level in the hierarchy, $n(n-1)/2$ questions must be answered. For a large hierarchy, the number of questions to be answered grows very large. As discussed in Section 1, we could simply require the decision maker to answer $n - 1$ questions by filling in the first column. However, why the first column and not the second? The redundancy in questioning which is an inherent part of the AHP is essential if reasonable estimates of priorities are to be obtained. However, we may be able to reduce the number of questions to something between n - 1 and n(n - 1)/2. The *incomplete pairwise comparison method* described in [13] proceeds as follows. Let us assume that the decision maker has completed at least the top row (and hence first column by reciprocity) of a pairwise comparison matrix or its equivalent. For the missing matrix entries a_{ij}, let us approximate their value by the ratio of the (yet unknown) weights w_i/w_j. For example, the following matrix has entry (1,3) missing:

$$C = \begin{bmatrix} 1 & 2 & w_1/w_3 \\ 1/2 & 1 & 2 \\ w_3/w_1 & 1/2 & 1 \end{bmatrix}.$$

If one computes the value of Cw, the following vector is obtained:

$$\begin{bmatrix} 2w_1 + 2w_2 \\ 1/2w_1 + w_2 + 2w_3 \\ 1/2w_1 + 2w_3 \end{bmatrix}.$$

Note that this vector could be obtained from multiplying the following matrix A by w; i.e., Aw = Cw:

$$A = \begin{bmatrix} 2 & 2 & 0 \\ 1/2 & 1 & 2 \\ 0 & 1/2 & 2 \end{bmatrix}.$$

Thus, we simply place a zero in the matrix when a question has not been answered, and add one to the diagonal for each missing entry in a row. Harker [13] has shown that the same theory and computational procedure for positive, reciprocal matrices holds for this *nonnegative, quasi-reciprocal matrix A*. Applying this computational procedure to the matrix A yields

$$w = (4/7,\ 2/7,\ 1/7)$$

$$\lambda_{max} = 3.$$

In summary, the incomplete comparison method allows one to reduce the effort involved in the elicitation of pairwise comparisons while at the same time allowing for the redundancy which is an important component of the AHP.

The second extension which is described in [13] involves nonlinear responses to the question: "compare i against j." The initial development of the AHP assumes that people respond to such a question with an estimate of the ratios of the relative weights; i.e.,

$$a_{ij} \approx w_i/w_j.$$

Table 23. Eigenvector for Various Powers α

α	w_1	w_2	w_3	w_4
0.1	0.999937	0.000063	0.000000	0.000000
0.2	0.991985	0.007898	0.000115	0.000002
0.4	0.907923	0.081015	0.009761	0.001301
0.8	0.694322	0.207404	0.071992	0.026282
1.0	0.618669	0.235323	0.100934	0.045074
2.0	0.436572	0.269252	0.176338	0.117839
5.0	0.321306	0.264825	0.223581	0.190288
10.0	0.284770	0.258532	0.237548	0.219150

However, evidence from psychology [20, p. 189] suggests that people may respond to various stimuli in a manner which is more consistent with a *power function*, i.e.,

$$a_{ij} \approx (w_i/w_j)^{\alpha},$$

where α is a positive scalar parameter. For values of $\alpha < 1$, the differences in the weights will tend to be exaggerated, while values of $\alpha > 1$ tend to reduce these differences. For example, consider the following matrix:

$$A = \begin{bmatrix} 1 & 5 & 6 & 7 \\ 1/5 & 1 & 4 & 6 \\ 1/6 & 1/4 & 1 & 4 \\ 1/7 & 1/6 & 1/4 & 1 \end{bmatrix} .$$

The weights for various values of α are shown in Table 23; these results confirm the above intuition. Computationally, these weights are computed precisely in the same manner as in Appendix A:

$$Aw^{\alpha} = \lambda_{max} w^{\alpha}.$$

Once one has computed $w^{\alpha} = (w_1^{\alpha}, w_2^{\alpha}, \ldots, w_n^{\alpha})$, the final weights can be computed as the αth root of each component of the eigenvector; see [13] for details.

The power function approach may be useful in pre-testing the decision maker through the use of physical estimation problems (e.g., weights, distance, brightness, etc.) to calibrate the value of α in order to obtain better estimates of the decision maker's preferences and to overcome either the "even keel" mentality in which no high numbers are ever assigned in the matrix or the "radical" mentality in which either 1 or 9 is assigned. This pre-testing of decision makers is a fruitful area for future research.

APPLICATIONS OF THE ANALYTIC HIERARCHY PROCESS: A CATEGORIZED, ANNOTATED BIBLIOGRAPHY

Bruce L. Golden
College of Business and Management
University of Maryland
College Park, Maryland 20742

Edward A. Wasil
Kogod College of Business Administration
American University
Washington, D.C. 20016

Doug E. Levy
Chesapeake Biological Laboratory
University of Maryland
Solomons, Maryland 20688

ABSTRACT

Since its introduction in the mid 1970s, the Analytic Hierarchy Process has been applied to many types of decision problems. In this paper, we identify more than 150 published papers that use the AHP to model diverse problems and we categorize each paper according to 29 application areas that range from conflict analysis to urban planning. In addition, we classify papers that combine the AHP with some traditional operations research techniques (e.g., linear programming) to analyze alternatives. Finally, in order to convey both the practicality and impact of this technique, we annotate 17 papers that either model important, real-world problems or apply the AHP in an interesting or unusual setting.

1. INTRODUCTION

Since its introduction by Saaty [10][1] over ten years ago, the Analytic Hierarchy Process has been applied in a wide variety of practical settings to model complex decision problems. The ability to rank decision alternatives based on both qualitative and quantitative factors using the AHP has led to many applications in such diverse areas as health care, politics, urban planning, and space exploration. The AHP has been used in ranking, selection, evaluation, optimization, and prediction decision problems. It has been combined with well-known operations research techniques, such as integer and linear programming, to form "hybrid" tools that can produce insightful results to difficult problems. AHPs wide-ranging appeal as a decision-analysis tool is reflected in the number and diversity of

---

[1] We cite papers, books, and software not appropriate for the bibliography with [ ] and list them in the references section, while papers listed in the bibliography are cited with ( ).

application papers that have already appeared in the literature. From 1976-1988 we have identified 153 published papers in 29 different application areas. Nearly as many papers have appeared in the literature on the theoretical aspects of the AHP, but we do not focus on these in this paper.

In the remainder of this paper, we categorize the large body of AHP applications in the form of several easy-to-use classification tables and we annotate 17 papers that convey the diversity, practicality, and impact of the AHP as a decision-modeling tool. The comprehensive listing of applied AHP paper citations is also included. For discussions on the theoretical aspects of the AHP, the papers by Saaty [10, 11] and Harker and Vargas [7] and the paper by Harker in this volume are especially informative. A well-written, practitioner's guide to the AHP is provided by Zahedi [20]. Four major books that present theory and applications are also available: **The Analytic Hierarchy Process** (Saaty [13]), **Decision Making for Leaders** (Saaty [12]), **The Logic of Priorities: Applications in Business, Energy, Health, and Transportation** (Saaty and Vargas [16]), and **Analytical Planning-The Organisation of Systems** (Saaty and Kearns [15]). Several college-level textbooks devote a chapter to the AHP, including **Thinking with Models** (Saaty and Alexander [14]), **Decision Making, Models and Algorithms** (Gass [4]) and **Quantitative Methods for Business** (Anderson, Sweeney, and Williams [1]).

Interested readers can also refer to the three special issues of academic journals listed below that are devoted entirely to the AHP.

- 1983 **Mathematics and Computers in Simulation** [19]
  This issue contains six articles that focus on the modeling of social decision processes in areas such as conflict resolution, energy policy planning, and health care.

- 1986 **Socio-Economic Planning Sciences** [6]
  A mix of theory and application in areas such as utility theory and multiobjective planning is presented. This issue contains thirteen articles and is guest edited by Patrick Harker.

- 1987 **Mathematical Modelling** [9]
  Three issues, guest edited by R.W. Saaty and L.G. Vargas, are devoted to theoretical developments and applications. Thirteen articles focus on applications in areas such as finance and marketing, macroeconomic forecasting, and legal case planning. Eleven articles examine theoretical issues in areas such as utility theory, consistency, and dependency.

Although not devoted entirely to the AHP, **Mathematical Modelling** [2] includes eight AHP papers covering theory and applications which were presented at the Fifth International Conference on Mathematical Modelling in Science and Technology.

We point out that the proliferation of AHP applications has been facilitated by the availability of microcomputer software. **Expert Choice** [3] is a popular AHP package for the IBM PC that models problems in an easy-to-use format. Hamalainen et al. [5]

describe the development of **The Analytic Manager** microcomputer software in Finland. Saaty [12] also provides interactive computer programs written in Basic, Fortran, and APL for users interested in coding the AHP on a micro or mainframe system.

Finally, the AHP "phenomenon" is not solely restricted to applications carried out in the United States. About 30% of the contributions listed in the bibliography at the end of this paper are by scholars outside the U.S. Along these lines, Liu and Xu [8] cite 38 papers and reports written in Chinese about the AHP. They report that since the introduction of the AHP to China in 1982 "...thousands of Chinese people have used the method to solve problems encountered in their work." Vachnadze and Markozashvili [18] report on the use of the AHP in the Soviet Union. Tone [17] has also published a book about the AHP written in Japanese.

2. DISCUSSION OF BIBLIOGRAPHY

The papers cited in the bibliography at the end of this paper form a comprehensive, up-to-date list of AHP applications (as of September 1, 1988). For the most part, these papers are published in academic and practitioner journals, proceedings of conferences available as books (i.e., books cataloged by the Library of Congress), and books focusing on the AHP. Technical reports, unpublished manuscripts, and uncataloged proceedings are not included. Our bibliography of application papers includes 153 citations. As far as we know, the only other substantial listing of AHP papers is contained in the recent survey by Zahedi [20]. About 50 of our 153 citations are also listed by Zahedi.

In examining the entire set of papers, we were impressed by the diverse nature of the applications. To convey this diversity, we identified 29 broad application areas, ranging from accounting and finance to transportation, and classified each of the papers according to the appropriate area. This classification scheme is presented in Table 1. In this table, papers can appear in more than one area. For example, the paper by Gholamnezhad and Saaty (48) appears in both the energy and long range planning categories.

We were also surprised by the number of applications in which the AHP is combined with a traditional operations research technique, such as linear programming, to evaluate alternatives. For example, the papers by Liberatore (74, 75) develop a detailed AHP scoring model that assigns priorities to R&D project proposals. The project priorities then become objective function coefficients in a 0-1 integer linear program (ILP) designed to allocate resources amongst the competing projects. The ILP maximizes total priority over all projects subject to budgetary and other constraints. In all, we identified about 40 papers that combine the AHP with 15 operations research and related techniques. This classification scheme is presented in Table 2.

Although most of the papers listed in the bibliography involve straightforward applications of the AHP (i.e., decision problems are modeled using a single hierarchy and priorities are

Table 1. Application Areas

Application	Paper Number in Bibliography
Accounting and Finance	16, 17, 28, 36, 62, 76, 91
Architecture and Design	110, 114, 117
Capital Investment	53, 147
Computers and Information Systems	10, 11, 13, 26, 130, 131, 141, 153
Conflict Analysis	1, 2, 3, 7, 9, 27, 47, 104, 113, 128, 140, 145
Decision Support	15, 38, 45, 55
Economics	60, 96, 98, 102, 125, 137, 138, 149
Energy	46, 47, 48, 54, 55, 61, 68, 69, 73, 85, 89, 108, 113, 116, 118, 119, 121, 144
Futures Research	65, 66, 67
Group Decision Making	4, 44, 52, 57, 63, 78, 79, 89, 90
Health Care	33, 35, 57, 86, 93, 135
Higher Education	5, 22, 29, 30, 71, 84, 122, 123, 136
Long Range Planning	37, 48, 54, 109, 112, 116, 123
Manufacturing and Production	12, 14, 42, 132, 133, 139, 148
Marketing	6, 18, 19, 135, 150, 151, 152
Military	9, 39, 41, 43, 58, 88, 92, 106 128
Optimization	8, 43, 49, 70, 80, 81, 82, 83, 94, 95, 105
Politics	97, 98, 99, 100, 106, 115, 120, 128
Portfolio Selection	20, 124
Public Sector and Legal Planning	34, 51, 52, 87
Regional and Urban Planning	21, 31, 32, 59
R & D Management	23, 44, 72, 74, 75, 78, 85
Resource Allocation	4, 5, 72, 74, 75, 121
Risk Analysis	53, 64, 108, 142
Sociology	101, 129
Space Exploration	23, 24
Sports and Games	25, 40, 50, 56, 126, 127
Surveys of Applications	77, 103, 107, 111, 143
Transportation	109, 112, 134, 146

developed for the various levels of the hierarchy), several papers rely on sophisticated AHP features to model problems. The sophisticated features include the **analytic network process, cost-benefit analysis,** and the **forward-backward process.** We briefly describe each feature below and refer the interested reader to Saaty [13] for more technical details and also to Table 3 which lists papers that use these features.

AHPs basic structure allows interdependencies only among elements at two adjacent levels of the hierarchy. To handle feedback between other levels, the hierarchical structure is abandoned and replaced by a network (analytic network process) where nodes correspond to elements at a level of the hierarchy. Arcs between nodes signify interdependencies among elements. A supermatrix is then used to establish the priorities of the elements taking into account the interdependencies.

To evaluate alternatives in a cost-benefit framework, two hierarchies, a benefits hierarchy and a costs hierarchy, can be constructed. In this setting, the ratio of overall benefit priority to overall cost priority for each alternative forms the basis of comparison.

The forward-backward process has been used to analyze future outcomes in planning and conflict resolution problems. In the forward process, decision makers hierarchically analyze possible future scenarios and determine which are the most likely to occur. A backward process hierarchy is then used to determine which policies decision makers should pursue in order to achieve the desired future scenarios. Insights gained from the backward process can then be undertaken, and so on. The two processes can thus be applied in an iterative manner until a stable future scenario is achieved.

3. ANNOTATIONS

In this section, we annotate 17 papers that illustrate the application of the AHP to important or interesting decision problems drawn from six areas: conflict analysis, comparing nonlinear programming codes, military OR, regional and urban planning, R & D management, and space exploration. Many of the annotated applications use sophisticated features (such as the forward-backward process) or combine the AHP with an operations research technique in an interesting way.

3.1 Conflict Analysis

One of T.L. Saaty's major research interests over the last ten years has been conflict analysis and resolution. He and others have written numerous articles applying the AHP to this area. In this section, we focus on three interesting and provocative conflict analysis studies.

Saaty et al. (128) examine President Carter's decision to rescue the U.S. hostages from Iran in 1980. They apply the AHP after the fact in order to shed light on the subjective factors that led to this decision. Arguing that military experts advised the president that the likelihood of moderate success was high,

Table 2. Operations Research and Related Techniques

Technique	Paper Number in Bibliography
Applied Probability	126, 127
Cross Impact Analysis	65, 66
Dynamic Programming	149
Fuzzy Sets	81
Game Theory	47, 113
Goal Programming	29, 43, 71, 95, 136
Input-Output Analysis	125, 137, 138
Integer Programming	23, 35, 71, 73, 74, 75
Linear Programming	88, 121
Multiobjective Optimization	8, 23, 32, 43, 69, 70, 94, 95, 136
Networks	136
Nonlinear Programming	23
Quadratic Programming	59
Time Series	31
Utility Theory	58, 60, 92, 128, 146

Table 3. AHP Features

Feature	Paper Number in Bibliography
Analytic Network Process	9, 27, 54, 55, 102, 108
Cost-Benefit Analysis	10, 12, 89, 98, 104, 108, 133, 139, 147, 148
Forward-Backward Process	1, 2, 3, 37, 47, 93, 140, 145

the authors' analysis reveals that Carter's political future was the most significant decision factor, followed by U.S. prestige. This is consistent with the resignation of Carter's secretary of state over the decision, since he would naturally place less weight on the president's political future than Carter himself. Results from applying multiattribute utility theory are also compared to the AHP results.

In Saaty (113), the AHP is applied to an analysis of the U.S. -OPEC oil conflict. First, a number of U.S. and OPEC strategies are identified and then ranked separately with respect to overall objectives. The priorities yielded by this process become the "intrinsic" values of the various strategies. Next, U.S. strategies are matched against specific OPEC strategies, and vice versa, to determine the "relative strength" values. A payoff matrix to the U.S. is then constructed using these intrinsic and relative strength values. A payoff matrix to OPEC is computed similarly. The author searches the payoff matrix for equilibrium values and offers real-world insights.

Alexander and Saaty (2) apply the forward-backward process to the analysis of the conflict in Northern Ireland. They first determine the most likely political outcome or future. Next, they test the stability of that outcome by perturbing various comparison matrices. The authors conclude that the predicted outcome is sufficiently robust.

3.2 Comparing Nonlinear Programming Codes

In the last twenty years, numerous studies have appeared in the literature comparing nonlinear programming codes. To evaluate the codes, researchers measure performance on objective criteria such as accuracy and efficiency, as well as subjective criteria such as ease of use. Combining both types of measurements so as to identify the best performing codes is extremely challenging. Lootsma (80, 81, 82, 83) provides a framework based on the AHP for comparing nonlinear programming codes. This framework is especially useful since it provides researchers with a way of comparing codes in the presence of failures (i.e., all codes do not solve all problems). Golden and Wasil (49) use Lootsma's method to compare microcomputer-based software packages that solve nonlinear programs and systems of simultaneous nonlinear equations.

3.3 Military OR

Mitchell and Bingham (88) describe a project carried out for the Canadian Department of National Defence in which they seek to maximize the benefits to be achieved from the repair and overhaul of Canadian Forces land-based equipment (ranging from battle tanks to laundry-washing machines) subject to resource and facilities limitations. AHP and linear programming are used in tandem in order to address this public sector allocation problem to the satisfaction of the key decision makers.

Gass (43) proposes the use of the AHP to help automate the generation of thousands of weights needed to arrive at acceptable solutions to large-scale linear goal programming models. The example given is a U.S. Army model which aids in managing the flow of personnel so as to best meet a desired end-strength goal over a 7-year planning horizon. AHP-derived weights serve as objective function coefficients in a large-scale, non-preemptive goal program.

Hannan et al. (58) review a model combining the AHP with multiattribute utility theory and its use by the U.S. Coast Guard to aid in the selection of auxiliary devices for icebreakers. Five auxiliary devices were considered and, as an outcome of this modeling exercise, the Coast Guard increased its use of the device which received the highest priority.

3.4 Regional and Urban Planning

An interesting problem central to public policy analyses is the prediction of population migration patterns. Cook et al. (31) present a new forecasting method that combines time series analysis with the AHP to predict intraurban migration patterns. Household population is forecast using Box-Jenkins methods and models are developed for the population share of each division encompassing a metropolitan area. AHP is then used to predict these population shares by evaluating the impacts of external factors (such as the amount of building activity in a division) that cannot be captured by the time series analysis. The statistical and AHP forecasts are then combined in an adjustment process that solves a simple linear program to produce each division's final share of the population. The AHP-adjusted forecasting method produced accurate results in predicting the 1979 household population for six divisions encompassing Portland, Oregon.

In a related paper, Harker (59) combines the AHP with a simple quadratic programming model to make predictions of interregional migration patterns. The approach uses the AHP to weight predictions based on the proximity of one region to another, quality of life factors, and economic factors to produce a single, composite prediction for each region.

3.5 R&D Management

The research and development (R&D) project selection decision involves allocating resources to a set of project proposals. A number of approaches such as 0-1 goal programming and multiattribute utility theory have been applied in the literature. Liberatore (74, 75) introduces a hierarchical-based framework to model this problem in which the AHP is linked with Lotus 1-2-3 to yield a project rating spreadsheet model that assigns priorities to each proposal. These priorities are then used as objective function coefficients in a 0-1 integer linear program that is designed to allocate limited organizational resources (such as funds and manpower) amongst the proposals.

3.6 Space Exploration

Bard (23) uses the AHP in conjunction with multiobjective optimization to determine the optimal level of subsystem automation for the space station currently being designed by the Johnson Space Center in Texas.

First, a multiobjective mathematical program is solved for a set of nondominated solutions. Next, these nondominated solutions are ranked using the AHP to produce the "optimal" level of automation with respect to monitoring, verification/calibration, fault management, extra vehicular activity, and docking functions. The alternatives here are rather general. In Bard's experiment, the most fully automated option received the highest priority due to the importance of "safety" issues.

In a companion paper, Bard (24) compares five specific alternatives for an on-orbit assembly station using the AHP. Sensitivity analysis is also performed.

4. CONCLUSIONS

In the last decade or so, the AHP has been applied to a multitude of diverse decision problems. Researchers and practitioners have used the AHP to model important real-world problems, to reach insightful decisions, and to offer provocative solutions to complex problems. In the next decade, we expect an even wider diffusion of use and the number of new applications and successful implementations to grow quite rapidly.

5. REFERENCES

1. D. Anderson, D. Sweeney, and T. Williams, **Quantitative Methods for Business**, West Publishing Company, St. Paul, Minnesota (1986).

2. X. Avula, G. Leitmann, C. Mote, and E. Rodin, editors, **Mathematical Modelling**, Special Issue on Mathematical Modelling in Science and Technology, **8** (1987).

3. **Expert Choice**, Decision Support Software, Inc., McLean, Virginia (1986).

4. S. Gass, **Decision Making, Models and Algorithms**, John Wiley & Sons, New York, New York (1985).

5. R. Hamalainen, T. Seppalainen, and J. Ruusunen, "A Microcomputer-Based Decision Support Tool and Its Application to a Complex Energy Decision Problem," in **Architecture Decision Support Systems and Knowledge-Based Systems: Special Topics**, Y. Chu, L. Haynes, L. Hoevel, A. Speckhard, E. Stohr, and R. Sprague, editors, Western Periodicals, North Hollywood, California (1986).

6. P. Harker, editor, **Socio-Economic Planning Sciences**, Special Issue on The Analytic Hierarchy Process, **20**, No. 6 (1986).

7. P. Harker, and L. Vargas, "The Theory of Ratio Scale Estimation: Saaty's Analytic Hierarchy Process," **Management Science, 33**, 1383-1403 (1987).

8. B. Liu and S. Xu, "Development of the Theory and Methodology of the Analytic Hierarchy Process and Its Applications in China," **Mathematical Modelling, 9**, 179-183 (1987).

9. R. Saaty and L. Vargas, editors, **Mathematical Modelling**, Special Issue on The Analytic Hierarchy Process: Theoretical Developments and Some Applications, **9**, No. 3-5 (1987).

10. T. Saaty, "A Scaling Method for Priorities in Hierarchical Structures," **Journal of Mathematical Psychology, 15**, 234-281 (1977).

11. T. Saaty, "Axiomatic Foundation of the Analytic Hierarchy Process," **Management Science, 32**, 841-855 (1986).

12. T. Saaty, **Decision Making for Leaders**, Lifetime Learning Publications, Belmont, California (1982).

13. T. Saaty, **The Analytic Hierarchy Process**, McGraw-Hill, New York, New York (1980).

14. T. Saaty and J. Alexander, **Thinking with Models**, Pergamon Press, Oxford, England (1981).

15. T. Saaty and K. Kearns, **Analytical Planning - The Organisation of Systems**, Pergamon Press, Oxford, England (1985).

16. T. Saaty and L. Vargas, **The Logic of Priorities: Applications in Business, Energy, Health, and Transportation**, Kluwer-Nijhoff, Boston, Massachusetts (1982).

17. K. Tone, **The Analytic Hierarchy Process**, in Japanese (1986).

18. R. Vachnadze and N. Markozashvili, "Some Applications of the Analytic Hierarchy Process," **Mathematical Modelling, 9**, 185-191 (1987).

19. R. Vichnevetsky, editor, **Mathematics and Computers in Simulation**, Special Issue on Modeling of Social Decision Processes, **25**, No. 2 (1983).

20. F. Zahedi, "The Analytic Hierarchy Process - A Survey of the Method and Its Applications," **Interfaces, 16**, No. 4, 96-108 (1986).

6. BIBLIOGRAPHY

1. Alexander, J.M. Priorities and preferences in conflict resolution. Mathematics and Computers in Simulation, 1983, 25(2), 108-119.

2. Alexander, J.M., & Saaty, T.L. Stability analysis of the forward-backward process: Northern Ireland case study. Behavioral Science, 1977, 22(6), 375-382.

3. Alexander, J.M., & Saaty, T.L. The forward and backward processes of conflict analysis. Behavioral Science, 1977, 22(2), 87-98.

4. Algie, J., Mallen, G., & Foster, W. Financial cutback decisions by priority scaling. Journal of Management Studies, 1983, 20(2), 233-260.

5. Arbel, A. A university budget problem: A priority-based approach. Socio-Economic Planning Sciences, 1983, 17(4), 181-189.

6. Arbel, A. Venturing into new technological markets. Mathematical Modelling, 1987, 9(3-5), 299-308.

7. Arbel, A., & Novik, N. U.S. pressure on Israel--likelihood and scope. Journal of Conflict Resolution, 1985, 29(2), 253-282.

8. Arbel, A., & Oren, S.S. Generating search directions in multiobjective linear programming using the analytic hierarchy process. Socio-Economic Planning Sciences, 1986, 20(6), 369-373.

9. Arbel, A., Saaty, T.L., & Vargas, L.G. Nuclear balance and the parity index: The role of intangibles in decisions. IEEE Transactions on Systems, Man, and Cybernetics, 1987, SMC17(5), 821-828.

10. Arbel, A., & Seidmann, A. Capacity planning, benchmarking and evaluation of small computer systems. European Journal of Operational Research, 1985, 22(3), 347-358.

11. Arbel, A., & Seidmann, A. Designing a multi-micro distributed accounting information system. Human Systems Management, 1985, 5(4), 238-291.

12. Arbel, A., & Seidmann, A. Performance evaluation of flexible manufacturing systems. IEEE Transactions on Systems, Man, and Cybernetics, 1984, SMC-14(4), 606-617.

13. Arbel, A., & Seidmann, A. Selecting a microcomputer for process control and data acquisition. IIE Transactions, 1984, 16(1), 73-80.

14. Arbel, A., & Shapira, Y. A decision framework for evaluating vacuum pumping technology. Journal of Vacuum Science and Technology, 1986, 4(2), 230-236.

15. Arbel, A., & Tong, R.M. On the generation of alternatives in decision analysis problems. OR: The Journal of the Operational Research Society, 1982, 33(4), 377-387.

16. Arrington, C.E., Hillison, W., & Jensen, R.E. An application of analytical hierarchy process to model expert judgments on analytical review procedures. Journal of Accounting Research, 1984, 22(1), 298-312.

17. Arrington, C.E., Jensen, R.E., & Tokutani, M. Scaling of corporate multivariate performance criteria: Subjective composition versus the analytic hierarchy process. Journal of Accounting and Public Policy, 1982, 1(2), 95-123.

18. Bahmani, N., & Blumberg, H. Consumer preference and reactive adaptation to a corporate solution of the over-the-counter medication dilemma - an analytic hierarchy process analysis. Mathematical Modelling, 1987, 9(3-5), 293-298.

19. Bahmani, N., Javalgi, R.G., & Blumburg, H. An application of the analytical hierarchy process for a consumer choice problem. Developments in Marketing Science, 1986, 9, 402-406.

20. Bahmani, N., Yamoah, D., Basseer, P., & Rezvani, F. Using the analytic hierarchy process to select investment in a heterogeneous environment. Mathematical Modelling, 1987, 8, 157-162.

21. Banai-Kashani, A.R. A paradigm for location and multidimensional problems in planning. Socio-Economic Planning Sciences, 1984, 18(3), 159-166.

22. Barbeau, E. Perron's result and a decision on admissions tests. Mathematics Magazine, 1986, 59(1), 12-22.

23. Bard, J.F. A multiobjective methodology for selecting subsystem automation options. Management Science, 1986, 32(12), 1628-1641.

24. Bard, J.F. Evaluating space station applications of automation and robotics. IEEE Transactions on Engineering Management, 1986, EM-33(2), 102-111.

25. Bartoszynski, R., & Puri, M.L. A note on predicting the results of chess championship matches. Behavioral Science, 1981, 26(1), 85-87.

26. Beck, M.P., & Lin, B.W. Selection of automated office systems: A case study. OMEGA, 1981, 9(2), 169-176.

27. Bennett, J.P., & Saaty, T.L. Terrorism: Patterns for negotiation; a case study using hierarchies and holarchies. In R.H. Kupperman & D.M. Trent (Eds.), Terrorism: Threat, Reality, Response, Stanford: Hoover Institution Press, 1979.

28. Blair, A.R., Nachtmann, R., Olson, J.E., & Saaty, T.L. Forecasting foreign exchange rates: An expert judgment approach. Socio-Economic Planning Sciences, 1987, 21(6), 363-369.

29. Brown, P.J.B. Trade-off and compromise in secondary school reorganization schemes: A goal programming approach. Papers of the Regional Science Association, 1987, 61, 115-130.

30. Canada, J.R., Frazelle, E.H., Koger, R.K., & MacCormac, E. How to make a career choice: The use of the analytic hierarchy process. Industrial Management, 1985, 27(5), 16-22.

31. Cook, T., Falchi, P., & Mariano, R. An urban allocation model combining time series and analytic hierarchical methods. Management Science, 1984, 30(2), 198-208.

32. Debeljak, C.J., Haimes, Y.Y., & Leach, M. Integration of the surrogate worth trade-off method and the analytic hierarchy process. Socio-Economic Planning Sciences, 1986, 20(6), 375-385.

33. Dolan, J.G. Clinical decision making using the analytic hierarchy process: Choice of antibiotic treatment for community-acquired pyelonephritis. Clinical Research, 1987, 35(3), 738.

34. Dorweiler, V.P. Legal case planning via the analytic hierarchy process. Mathematical Modelling, 1987, 9(3-5), 251-261.

35. Dougherty, J.J., & Saaty, T.L. Optimum determination of hospital requirements. In T.L. Saaty & L.G. Vargas (Eds.), The Logic of Priorities: Applications in Business, Energy, Health, and Transportation, Boston: Kluwer-Nijhoff Publishing, 1982.

36. Driscoll, D.A., Lin, W.T., & Watkins, P.R. Cost-volume-profit analysis under uncertainty: A synthesis and framework for evaluation. Journal of Accounting Literature, 1984, 3, 85-115.

37. Emshoff, J.R., & Saaty, T.L. Applications of the analytic hierarchy process to long range planning processes. European Journal of Operational Research, 1982, 10(2), 131-143.

38. Forman, E.H. Decision support for executive decision makers. Information Strategy: The Executive's Journal, 1985, 1(4), 4-14.

39. Forman, E.H. Decision support for the evaluation of the strategic defense initiative (star wars). In S.I. Gass, H.J. Greenberg, K.L. Hoffman, & R.W. Langley (Eds.), Impacts of Microcomputers on Operations Research, Amsterdam: Elsevier Publishing, 1986.

40. Forman, E.H. Relative vs. absolute worth. Mathematical Modelling, 1987, 9(3-5), 195-202.

41. Forman, E.H. Sensitivity investigation for a decision analysis of the strategic defense initiative (star wars). Mathematical Modelling, 1987, 8, 133-138.

42. Frazelle, E. Suggested techniques enable multi-criteria evaluation of material handling alternatives. Industrial Engineering, 1985, 17(2), 42-48.

43. Gass, S.I. A process for determining priorities and weights for large-scale linear goal programmes. OR: The Journal of the Operational Research Society, 1986, 37(8), 779-785.

44. Gear, T.E., Lockett, A.G., & Muhlemann, A.P. A unified approach to the acquisition of subjective data in R&D. IEEE Transactions on Engineering Management, 1982, EM-29(1), 11-19.

45. Gerrish, W. Pain in creative writing: Quick relief using the analytic hierarchy process. Mathematics and Computers in Simulation, 1983, 25(2), 120-124.

46. Gholamnezhad, A.H. A new model in the energy policy planning. Mathematics and Computers in Simulation, 1983, 25(2), 125-134.

47. Gholamnezhad, A.H. Critical choices for OPEC members and the United States. Journal of Conflict Resolution, 1981, 25(1), 115-143.

48. Gholamnezhad, A.H., & Saaty, T.L. A desired energy mix for the United States in the year 2000: An analytic hierarchy approach. International Journal of Policy Analysis and Information Systems, 1982, 6(1), 47-64.

49. Golden, B.L., & Wasil, E.A. Nonlinear programming on a microcomputer. Computers & Operations Research, 1986, 13(2/3), 149-166.

50. Golden, B.L., & Wasil, E.A. Ranking outstanding sports records. Interfaces, 1987, 17(5), 32-42.

51. Grizzle, G.A. Pay for performance: Can the analytic hierarchy process hasten the day in the public sector? Mathematical Modelling, 1987, 9(3-5), 245-250.

52. Grizzle, G.A. Priority-setting methods for plural policymaking bodies. Administration and Society, 1985, 17(3), 331-359.

53. Hachadorian, G.E. Hierarchical determination of the risk of forced divestment to foreign owned enterprises in LDC's. Mathematical Modelling, 1987, 8, 163-166.

54. Hamalainen, R.P., & Seppalainen, T.O. The analytic network process in energy policy planning. Socio-Economic Planning Sciences, 1986, 20(6), 399-405.

55. Hamalainen, R.P., Seppalainen, T.O., & Ruusunen, J. A microcomputer-based decision support tool and its application to a complex energy decision problem. In Y. Chu, L. Haynes, L.W. Hoevel, A. Speckhard, E.A. Stohr, & R.H. Sprague (Eds.), Architecture, Decision Support Systems and Knowledge-Based Systems: Special Topics, North Hollywood: Western Periodicals, 1986.

56. Hannan, E.L. An eigenvalue method for evaluating contestants. Computers & Operations Research, 1983 10(1), 41-46.

57. Hannan, E.L., O'Donnell, J., & Freedland, T. A priority assignment model for standards and conditions in a long term care survey. Socio-Economic Planning Sciences, 1981, 15(6), 277-289.

58. Hannan, E.L., Smith, J.A., & Gilbert, G.R. A multiattribute decision-making approach to the selection of an auxiliary device for icebreakers. Decision Sciences, 1983, 14(2), 240-252.

59. Harker, P.T. The use of expert judgements in predicting interregional migration patterns: An analytic hierarchy approach. Geographical Analysis, 1986, 18(1), 62-80.

60. Hughes, W.R. Deriving utilities using the analytic hierarchy process. Socio-Economic Planning Sciences, 1986, 20(6), 393-395.

61. Hussien, F.M., Obeid, M.A., & El-Malahy, K.S. Site selection of a dual purpose nuclear power plant in Saudi Arabia. Nuclear Technology, 1987, 79(3), 311-321.

62. Jensen, R.E. A dynamic analytic hierarchy process analysis of capital budgeting under stochastic inflation rates and risk premiums. Advances in Financial Planning and Forecasting, 1987, 2, 269-302.

63. Jensen, R.E. Comparison of consensus methods for priority ranking problems. Decision Sciences, 1986, 17(2), 195-211.

64. Jensen, R.E. International investment risk analysis: Extensions for multinational corporation capital budgeting models. Mathematical Modelling, 1987, 9(3-5), 265-284.

65. Jensen, R.E. Reporting of management forecasts: An eigenvector model for elicitation and review of forecasts. Decision Sciences, 1982, 13(1), 15-37.

66. Jensen, R.E. Scenario probability scaling: An eigenvector analysis of elicited scenario odds ratios. Futures, 1981, 13(6), 489-498.

67. Jensen, R.E., & Spencer, R.W. Matrix scaling of subjective probabilities of economic forecasts. Economic Letters, 1986, 20(3), 221-225.

68. Johnson, C.R., & Hihn, J.M. Prioritization of research projects. In J. Silverman (Ed.), Energy Storage: A Vital Element in Mankind's Quest for Survival and Progress, Oxford: Pergamon Press, 1980.

69. Kok, M., & Lootsma, F.A. Pairwise-comparison methods in multiple objective programming, with applications in a long-term energy-planning model. European Journal of Operational Research, 1985, 22(1), 44-55.

70. Korhonen, P. The specification of a reference direction using the analytic hierarchy process. Mathematical Modelling, 1987, 9(3-5), 361-368.

71. Kwak, N.K., & Diminnie, C.B. A goal programming model for allocating operating budgets of academic units. Socio-Economic Planning Sciences, 1987, 21(5), 333-339.

72. Lauro, G.L., & Vepsalainen, A.P.J. Assessing technology portfolios for contract competition: An analytic hierarchy process approach. Socio-Economic Planning Sciences, 1986, 20(6), 407-415.

73. Legrady, K., Lootsma, F.A., Meisner, J., & Schellemans, F. Multicriteria decision analysis to aid budget allocation. In M. Grauer & A.P. Weirzbicki (Eds.), Interactive Decision Analysis, Berlin: Springer-Verlag, 1984.

74. Liberatore, M.J. An extension of the analytic hierarchy process for industrial R&D project selection and resource allocation. IEEE Transactions on Engineering Management, 1987, EM-34(1), 12-18.

75. Liberatore, M.J. R&D project selection. Telematics and Informatics, 1987, 3(4), 289-300.

76. Lin, W.T., Mock, T., & Wright, A. The use of the analytic hierarchy process as an aid in planning the nature and extent of audit procedures. Auditing: A Journal of Practice & Theory, 1984, 4(1), 89-99.

77. Liu, B., & Xu, S. Development of the theory and methodology of the analytic hierarchy process and its applications in China. Mathematical Modelling, 1987, 9(3-5), 179-183.

78. Lockett, A.G., Hetherington, B., Yallup, P., Stratford, M., & Cox, B. Modelling a research portfolio using AHP: A group decision process. R&D Management, 1986, 16(2), 151-160.

79. Lockett, A.G., Muhlemann, A.P., & Gear, A.E. Group decision making and multiple criteria--a documented application. In C.L. Hwang & K. Yoon (Eds.), Multiple Attribute Decision Making - Methods and Applications: A State-of-the-Art Survey, Berlin: Springer-Verlag, 1981.

80. Lootsma, F.A. Performance evaluation of nonlinear optimization methods via multi-criteria decision analysis and via linear model analysis. In M.J.D. Powell (Ed.), Nonlinear Optimization 1981, New York: Academic Press, 1982.

81. Lootsma, F.A. Performance evaluation of nonlinear optimization methods via pairwise comparison and fuzzy numbers. Mathematical Programming, 1985, 33, 93-114.

82. Lootsma, F.A. Performance evaluation of non-linear programming codes from the viewpoint of a decision maker. In I.D. Fosdick (Ed.), Performance Evaluation of Numerical Software, Amsterdam: North-Holland Publishing, 1979.

83. Lootsma, F.A. Ranking of non-linear optimization codes according to efficiency and robustness. In L. Collatz, G. Meinardus, & W. Wetterling (Eds.), Konstruktive Methoden der Finiten Nichtlinearen Optimierung, Boston: Birkhauser Publishing, 1980.

84. Lootsma, F.A. Saaty's priority theory and the nomination of a senior professor in operations research. European Journal of Operational Research, 1980, 4(6), 380-388.

85. Lootsma, F.A., Meisner, J., & Schellemans, F. Multi-criteria decision analysis as an aid to the strategic planning of energy R&D. European Journal of Operational Research, 1986, 25(2), 216-234.

86. Lusk, E.J. Analysis of hospital capital decision alternatives: A priority assignment model. OR: The Journal of the Operational Research Society, 1979, 30(5), 439-448.

87. MacCormac, E.R. Values and technology: How to introduce ethical and human values into public policy decisions. Research in Philosophy and Technology, 1983, 6, 143-155.

88. Mitchell, K.H., & Bingham, G. Maximizing the benefits of Canadian forces equipment overhaul programs using multi-objective optimization. INFOR: The Canadian Journal of Operational Research, 1986, 24(4), 251-264.

89. Mitchell, K.H., & Soye, M.P. Measuring the intangibles in social decisions: Assessing benefits and costs of energy policy options. Mathematics and Computers in Simulation, 1983, 25(2), 135-145.

90. Mitroff, I.I., Emshoff, J.R., & Kilmann, R.H. Assumptional analysis: A methodology for strategic problem solving. Management Science, 1979, 25(6), 583-593.

91. Neim, A.J., & Teed, S.A. Assessing inherent risk in EDP applications: An instrumental addition to the auditor's toolbox. CA Magazine, 1986, 119(2), 42-48.

92. Ng, K.Y.K. On the choice of a military transport aircraft fleet via worth assessment procedure: A case study. IEEE Transactions on Systems, Man, and Cybernetics, 1980, SMC-10(12), 894-899.

93. Odynocki, B. The forward and backward processes in health policy planning. Mathematics and Computers in Simulation, 1983, 25(2), 146-155.

94. Olson, D.L. Opportunities and limitations of AHP in multiobjective programming. Mathematical and Computer Modelling, 1988, 11, 206-209.

95. Olson, D.L., Venkataramanan, M., & Mote, J.L. A technique using analytical hierarchy process in multiobjective planning models. Socio-Economic Planning Sciences, 1986, 20(6), 361-368.

96. Peniwati, K., & Hsiao, T. Ranking countries according to economic, social, and political indicators. Mathematical Modelling, 1987, 9(3-5), 203-209.

97. Rabinowitz, G. Some comments on measuring world influence. Journal of Peace Science, 1976, 2(1), 49-55.

98. Ramanujam, V., & Saaty, T.L. Technological choice in less developed countries: An analytic hierarchy approach. Technological Forecasting and Social Change, 1981, 19(1), 81-98.

99. Rapoport, A., Felsenthal, D.S., & Maoz, Z. Microcosms and macrocosms: Seat allocation in proportional representation systems. Theory and Decision, 1988, 24, 11-33.

100. Rapoport, A., Felsenthal, D.S., & Maoz, Z. Proportional representation: An empirical evaluation of single-stage, non-ranked voting procedures. Public Choice, 1988, 59(2), 151-165.

101. Saaty, T.L. Absolute and relative measurement with the AHP; the most livable cities in the United States. Socio-Economic Planning Sciences, 1986, 20(6), 327-331.

102. Saaty, T.L. A new macroeconomic forecasting and policy evaluation method using the analytic hierarchy process. Mathematical Modelling, 1987, 9(3-5), 219-231.

103. Saaty, T.L. Applications of analytical hierarchies. Mathematics and Computers in Simulation, 1979, 21(1), 1-20.

104. Saaty, T.L. Conflict resolution and the Falkland Islands invasions. Interfaces, 1983, 13(6), 68-83.

105. Saaty, T.L. Exploring optimization through hierarchies and ratio scales. Socio-Economic Planning Sciences, 1986, 20(6), 355-360.

106. Saaty, T.L. Impact of disarmament nuclear package reductions. In R. Avenhaus & R.K. Huser (Eds.), Quantitative Assessment in Arms Control, New York: Plenum Press, 1983.

107. Saaty, T.L. Priority setting in complex problems. IEEE Transactions on Engineering Management, 1983, EM-30(3), 140-155.

108. Saaty, T.L. Risk - it's priority and probability: The analytic hierarchy process. Risk Analysis, 1987, 7(2), 159-172.

109. Saaty, T.L. Scenarios and priorities in transport planning: Application to the Sudan. Transportation Research, 1977, 11(5), 343-350.

110. Saaty, T.L. The analytic hierarchy process: A new approach to deal with fuzziness in architecture. Architectural Science Review, 1982, 25(3), 64-69.

111. Saaty, T.L. The analytic hierarchy process: Decision making in complex environments. In R. Avenhaus & R.K. Huser (Eds.), Quantitative Assessment in Arms Control, New York: Plenum Press, 1983.

112. Saaty, T.L. The Sudan transport study. Interfaces, 1977, 8(1), 37-57.

113. Saaty, T.L. The US-OPEC energy conflict--the payoff matrix by the analytic hierarchy process. International Journal of Game Theory, 1979, 8(4), 225-234.

114. Saaty, T.L., & Beltran, M.H. Architectural design by the analytic hierarchy process. Design Methods and Theories, 1980, 14(3/4), 124-134.

115. Saaty, T.L., & Bennett, J.P. A theory of analytical hierarchies applied to political candidacy. Behavioral Science, 1977, 22(4), 237-245.

116. Saaty, T.L., & Bennett, J.P. XYZ research institute: Planning resource allocation under uncertainty. In T.L. Saaty & L.G. Vargas, The Logic of Priorities: Applications in Business, Energy, Health, and Transportation, Boston: Kluwer-Nijhoff Publishing, 1982.

117. Saaty, T.L., & Erdener, E. A new approach to performance measurement: The analytic hierarchy process. Design Methods and Theories, 1979, 13(2), 64-72.

118. Saaty, T.L., & Gholamnezhad, A.H. High-level nuclear waste management: Analysis of options. Environment and Planning B, 1982, 9(2), 181-196.

119. Saaty, T.L., & Gholamnezhad, A.H. Oil prices: 1985-1990. Energy Systems and Policy, 1981, 5(4), 303-318.

120. Saaty, T.L., & Khouja, M.W. A measure of world influence. Journal of Peace Science, 1976, 2(1), 31-48.

121. Saaty, T.L., & Mariano, R.S. Rationing energy to industries: Priorities and input-output dependence. Energy Systems and Policy, 1979, 3(1), 85-111.

122. Saaty, T.L., & Ramanujam, V. An objective approach to faculty promotion and tenure by the analytic hierarchy process. Research in Higher Education, 1983, 18(3), 311-331.

123. Saaty, T.L., & Rogers, P.C. Higher education in the United States (1985-2000): Scenario construction using a hierarchical framework with eigenvector weighting. Socio-Economic Planning Sciences, 1976, 10(6), 251-263.

124. Saaty, T.L., Rogers, P.C., & Pell, R. Portfolio selection through hierarchies. The Journal of Portfolio Management, 1980, 6(3), 16-21.

125. Saaty, T.L., & Vargas, L.G. Estimating technological coefficients by the analytic hierarchy process. Socio-Economic Planning Sciences, 1979, 13(6), 333-336.

126. Saaty, T.L., & Vargas, L.G. Hierarchical analysis of behavior in competition: Prediction in chess. Behavioral Science, 1980, 25(3), 180-191.

127. Saaty, T.L., & Vargas, L.G. Modeling behavior in competition: The analytic hierarchy process. Applied Mathematics and Computation, 1985, 16(1), 49-92.

128. Saaty, T.L., Vargas, L.G., & Barzilay, A. High-level decisions: A lesson from the Iran hostage rescue operation. Decision Sciences, 1982, 13(2), 185-206.

129. Saaty, T.L., & Wong, M.M. Projecting average family size in rural India by the analytic hierarchy process. Journal of Mathematical Sociology, 1983, 9(3), 181-209.

130. Seidmann, A., & Arbel, A. An analytic approach for planning computerized office systems. OMEGA, 1983, 11(6), 607-617.

131. Seidmann, A., & Arbel, A. Microcomputer selection process for organizational information management. Information and Management, 1984, 7(6), 317-329.

132. Seidmann, A., Arbel, A., & Shapira, R. A two-phase analytic approach to robotic system design. Robotics & Computer-Integrated Manufacturing, 1984, 1(2), 181-190.

133. Shapira, Y., & Arbel, A. A systematic approach for selecting pumping technology: Ultrahigh vacuum application. Journal of Vacuum Science and Technology, 1987, 5(4), 2526-2529.

134. Sharp, J.A. Haulier selection - an application of the analytic hierarchy process. OR: The Journal of the Operational Research Society, 1987, 38(4), 319-328.

135. Schwartz, R.G., & Oren, S. Using analytic hierarchies for consumer research and market modelling. Mathematical and Computer Modelling, 1988, 11, 266-271.

136. Sinuany-Stern, Z. A network optimization model for budget allocation in a multi-campus university. OR: The Journal of the Operational Research Society, 1984, 35(8), 749-757.

137. Steenge, A.E. Consistency and composite numeraires in joint production input-output analysis: An application of ideas of T.L. Saaty. Mathematical Modelling, 1987, 9(3-5), 233-241.

138. Steenge, A.E. Saaty's consistency analysis: An application to problems in static and dynamic input-output models. Socio-Economic Planning Sciences, 1986, 20(3), 173-180.

139. Sullivan, W.G. Models IEs can use to include strategic, non-monetary factors in automation decisions. Industrial Engineering, 1986, 18(3), 42-50.

140. Tarbell, D.S., & Saaty, T.L. The conflict in South Africa: Directed or chaotic. Journal of Peace Science, 1980, 4(2), 151-168.

141. Toshtzar, M. Multi-criteria decision making approach to computer software evaluation: Application of the analytic hierarchy process. Mathematical and Computer Modelling, 1988, 11, 276-281.

142. Tummala, V.M.R., Smith, J.R., & Uppuluri, V.R.R. Evaluating commercial credit decisions by analytic hierarchy process. ASCI Journal of Management, 1982/83, 12(1/2), 48-56.

143. Vachnadze, R.G., & Markozashvili, N.I. Some applications of the analytic hierarchy process. Mathematical Modelling, 1987, 9(3-5), 185-191.

144. Vargas, L.G. Prediction and the analytic hierarchy process. Mathematics and Computers in Simulation, 1983, 25(2), 156-167.

145. Vargas, L.G. Prospects for the Middle East: Is a peaceful settlement attainable? European Journal of Operational Research, 1983, 14(2), 169-192.

146. Vargas, L.G., & Dougherty, J.J. The analytic hierarchy process and multicriterion decision making. American Journal of Mathematical and Management Sciences, 1982, 2(1), 59-92.

147. Vargas, L.G., & Saaty, T.L. Financial and intangible factors in fleet lease or buy decision. Industrial Marketing Management, 1981, 10(1), 1-10.

148. Varney, M.S., Sullivan, W.G., & Cochran, J.K. Justification of flexible manufacturing systems with the analytical hierarchy process. In C. Savage (Ed.), Focus on the Future: Proceedings of the 1985 Industrial Engineering Show, Atlanta: Industrial Engineering and Management Press, 1985.

149. Weiss, E.N. Using the analytic hierarchy process in a dynamic environment. Mathematical Modelling, 1987, 9(3-5), 211-216.

150. Wind, Y. An analytic hierarchy process based approach to the design and evaluation of a marketing driven business and corporate strategy. Mathematical Modelling, 1987, 9(3-5), 285-291.

151. Wind, Y., & Douglas, S. International portfolio analysis and strategy: The challenge of the 80s. Journal of International Business Studies, 1981, 12(2), 69-82.

152. Wind, Y., & Saaty, T.L. Marketing applications of the analytic hierarchy process. Management Science, 1980, 26(7), 641-658.

153. Zahedi, F. Data-base management system evaluation and selection decisions. Decision Sciences, 1985, 16(1), 91-109.

GROUP DECISION MAKING AND THE AHP

Thomas L. Saaty
Graduate School of Business
University of Pittsburgh
Pittsburgh, Pennsylvania 15260

ABSTRACT

This paper focuses on the application of the Analytic Hierarchy Process in a group setting. In particular, we present observations and suggestions that are intended to help in the planning and execution of a group decision-making effort in which AHP plays a major role.

1. INTRODUCTION

In the last few years, several articles in the management science literature (e.g., DeSanctis and Gallupe [2] and Huber and McDaniel [5]) have pointed to the following trend in decision making: Organizational decisions are much more technically and politically complex and require frequent group decision-making meetings. Decisions must be reached quickly, usually with greater participation of low-level or staff personnel than in the past. Many of these articles also focus on the development of computer-based systems that support the formulation and solution of unstructured decision problems by a group (i.e., a group decision support system or GDSS). One of the decision-aiding tools that can be used as part of a GDSS to help promote effective group interaction and participation is the Analytic Hierarchy Process. Our goal in this paper is to report on our experiences over the last ten years or so in using the AHP in a group setting. In particular, we provide guidance on how a management science practitioner might structure a group decision-making effort in which the AHP is the central analysis tool. The discussion focuses on three key areas: (1) assembling the group, (2) running the decision-making session, and (3) implementing the results.

2. ASSEMBLING THE GROUP

The inherent complexity and uncertainty surrounding an organization's major problems usually necessitates the participation of many individuals in the decision-making process. In some cases, the composition of the group is fixed (such as the National Security Council advising the President), while in others, it is necessary to select a mix of "actors" to form the decision-making group (such as choosing a panel to investigate the Challenger disaster). The latter selection process requires specifying the number of experts, non-experts, staff personnel, and upper-level managers to participate, as well as choosing the appropriate individuals. This process can be difficult and time-consuming for many reasons.

Decision makers who are considered "powerful" members of the organization might refuse to participate in the process. These members are aware that their level of control and influence might be diminished in a group setting. They fear that the social and interactive nature of the group process might dilute their power and ability to direct policy within the organization. However, if powerful individuals actively participate, they are likely to strongly influence the process with respect to their preferences. In contrast, results generated by a group that consists solely of "low-level" managers with little power may not be useful. The danger is that the powerful managers will implement their own preferred solution without taking into account the valuable opinions and observations of others in the organization. We point out that participants who are considered "experts" can be especially troublesome. They may have strong ideas on the appropriate course of action and may not be easily swayed in their assessments.

One way of dealing with the "power differential" problem is to assemble a group of participants that have equal responsibility and stature within the organization. Collectively, these individuals can be treated as a decision-making "subgroup" that could help formulate and solve a part of the problem with which they are most knowledgeable. They could also contribute to discussions that involve higher or lower levels of management. This can be viewed as a sort of "shared" decision-making responsibility in which high-level management cooperates with subordinates. It has been our experience that high-level management often depends on low-level employees to gather the appropriate information on which to base their decisions. In this regard, the use of the AHP can serve to facilitate this data gathering. The AHP helps expose various levels of management to a broad range of information, views, and arguments.

3. RUNNING THE DECISION-MAKING SESSION

After the group has been chosen, the members should begin preparing for the decision-making session by formalizing their agenda, structuring the allowable interactions between participants, and clearly defining the purpose of the session in advance. They can seek answers to several questions (such as the ones listed below) that are designed to establish the ground rules for the session.

* Is the purpose of the session simply to improve the group's understanding of the important problem?
* Or is the purpose to reach a final solution to the problem?.
* Are the participants committed to generating and implementing a final solution?
* What is the best way to combine the judgments of the participants on various issues in order to produce a single group judgment?

3.1 Constructing the Hierarchy

Since the AHP is the specified decision-aiding tool, the group must first construct an appropriate decision hierarchy that reflects the problem under study. Most groups are willing to accept the basic hierarchical structure of the AHP as a rational way of modeling their problem. It allows a group with widely-varying perspectives to decompose a complex problem into its most basic and important components.

Debate and discussion among the participants can be used to generate a hierarchy that adequately represents the problem. Each member presents his own definitions, arguments, and viewpoints about factors at each level of the hierarchy. Of course, reaching agreement about the overall hierarchical structure of the decision problem may prove to be quite difficult. The hierarchy may be unacceptable to some participants who have already "made up their minds" about the final outcome. Still others may be unwilling to have one of their key decision factors relegated to a lower level in the hierarchy. For example, consider a group that is trying to decide on the best course of action to combat illegal drugs in a city. The police department's representative might argue that crime is one of the key decision factors and favors placing it near the top of the hierarchy. Another member might take a broader view and talk about the political, social, and economic factors. Crime then appears as one of the sub-criteria of the social factor. Of course, the police now feel that the importance of crime has been substantially diminished in this structure.

We point out that, in our experience, when the group has agreed to use the AHP in advance of the first meeting, the initial group interaction exhibits form and direction. When the judgment process begins in the next step, the hierarchical structure helps the group focus on only one aspect of the decision problem at a time.

3.2 Getting the Group to Agree

Once the group has agreed on the hierarchy, it must then generate entries for the pairwise comparison matrices at each level. There are two ways to generate these entries: (1) consensus vote and (2) individual judgments.

The first method requires that the group reach a consensus agreement on each entry in a matrix. Consider the hierarchy and matrix shown in Figure 1. At the first level, the group must pairwise compare the three decision factors and reach an agreement on each a_{ij} entry in the matrix. Of course, a considerable amount of discussion (and initial disagreement) among the participants might be required to produce this number. Additionally, if the hierarchy is large, this step could require a significant number of comparisons that are usually tedious and time-consuming. Recent papers by Harker [3,4] develop the **incomplete pairwise comparison method** that can be used to reduce the number of comparisons in a large hierarchy. The paper by Harker that appears in this volume also describes this technique.

Figure 1. Consensus Vote

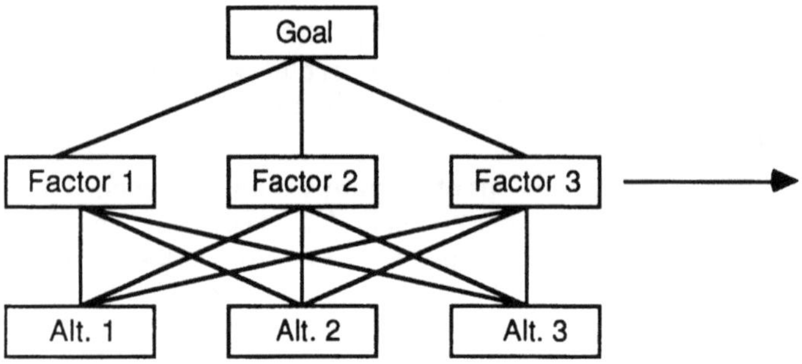

$$\begin{bmatrix} 1 & a_{12} & a_{13} \\ & 1 & a_{23} \\ & & 1 \end{bmatrix}$$

Figure 2. Combined Judgments

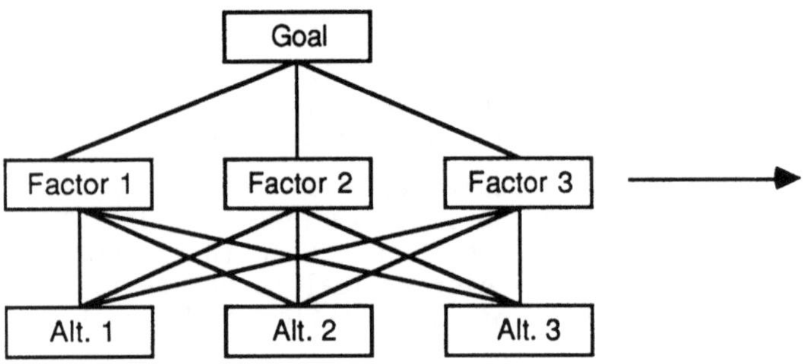

$$\begin{bmatrix} 1 & a_{12} & a_{13} \\ & 1 & a_{23} \\ & & 1 \end{bmatrix}$$

	Decision Maker			
	1	**2**	**. . .**	**N**
Judgments	a^1_{12}	a^2_{12}	. . .	a^N_{12}

Combined Judgments

$$a_{12} = [\, a^1_{12} \times a^2_{12} \times \ldots \times a^N_{12} \,]^{1/N}$$

Now suppose that there is substantial disagreement amongst the participants concerning an entry in the matrix. The AHP allows each decision maker to specify a value and then combine individual judgments as follows: Use the geometric mean of the individual judgments to obtain the group judgment for each pairwise comparison (see Figure 2). As shown by Aczel and Saaty [1], the geometric mean is the uniquely appropriate rule for combining judgments in the AHP because it preserves the reciprocal property in the combined pairwise comparison matrix.

An interesting question now arises: How well does an individual's set of judgments correspond with the group's set of judgments (as determined by the geometric mean)? We can give an answer to this question by examining the consistency measure of the AHP.

Let (a_{ij}) be an n x n pairwise comparison matrix formed by taking the geometric mean of the individual judgments and let $(w_1, w_2, \ldots, w_n)$ be the weights derived from this matrix. Then the expression

$$\sum_{i,j} a_{ij}w_j/w_i - n^2$$

gives the relative departure of (a_{ij}) from consistency. This follows from the fact that under perfect consistency we would have $a_{ij}w_j/w_i = 1$. Since $Aw = \lambda_{max}\, w$, we have

$$\sum_{j} a_{ij}w_j = \lambda_{max}\, w_i$$

or

$$\sum_{i,j} a_{ij}w_j/w_i = n\lambda_{max}.$$

If we take the deviation from n^2 (what we'd get with perfect consistency) and then divide by n^2, we obtain an index of relative departure from consistency:

$$(n\lambda_{max} - n^2)/n^2 = (\lambda_{max} - n)/n.$$

The consistency ratio (C.R.) is computed in the usual way by taking the ratio of the above expression to the random index. If the C.R. is less than 0.10, then the group judgment is consistent.

In the same way, we can determine the relative departure of an individual from the group judgment. If $(a_{ij}^{(k)})$ is the matrix of judgments of the kth individual and $(w_1, w_2, \ldots, w_n)$ are the weights calculated from the group pairwise comparison matrix (a_{ij}), then

$$\sum_{i,j} a_{ij}^{(k)} w_j/w_i - n^2$$

gives the relative departure of the kth individual's judgments from those of the group. The C.R. is computed in the usual way. For the individual to be compatible with the group, this measure should not exceed 0.10. This result enables users to identify individuals and subgroups that are at odds with the group as a whole.

In using the AHP, there have been times when participants felt that the judgmental process taxed their decision-making abilities. For this reason, a shorter procedure has been developed to elicit judgments by questionnaire.

In general, given n factors (or alternatives), n(n - 1)/2 judgments are required to generate the pairwise comparison matrix entries. As n increases, the work load necessary to elicit judgments also increases.

To diminish the work required to fill out a questionnaire, we present the following procedure that reduces the number of judgments from n(n - 1)/2 to n - 1. These n - 1 judgments form a spanning tree over the factors from which the entire comparison matrix can be constructed. Thus, by generating random spanning trees, one for each decision maker, we can capture the diversity of opinions within a group. Each group member will make n - 1 comparisons of possibly different factors or alternatives. The resulting n - 1 judgments are used to generate a consistent comparison matrix. All matrices are then combined into a single matrix using the geometric mean rule.

Next, we place the n - 1 judgments of each decision maker in another matrix. If this is done for all participants and the geometric mean is used when several decision makers answer questions corresponding to the same pairwise comparison, the resulting matrix may not have all its entries filled. When this is the case, the blank entries must be filled in with values from the consistent matrix that we constructed using the judgments of all participants. This matrix is used to estimate the relative priorities of the alternatives or factors and the consistency of the group. The process is then repeated for each criterion in the hierarchy.

3.3 Inequalities of Power

Factors such as personal charisma, perceived intelligence and expertise, and size and strength of an outside constituency can make some participants in a group decision-making effort "more powerful" than others. A key question is: Are such individuals willing to participate in a process in which strong and weak members have "equal" voices and are they willing to abide by the outcome of the group effort?

To help deal with the inequalities of power, the group at the outset of the process can develop weights for the importance of each participant. These weights can be used to emphasize judgmental contributions or to increase voting power. If the AHP is used in an organization where the formal ranks of participants are known and relative power can be assessed, then developing the weights is easy and natural. Developing these weights is difficult in groups where the power of each participant is nearly equal. Furthermore, in newly formed groups or ad hoc groups, it may not be possible or practical to gather enough information about the participants to make informed judgments about relative power. In such situations, the AHP can be a valuable interactive learning tool before powerful subgroups or individuals have been able to assert themselves.

An alternative to developing weights for each participant is simply to encourage the expectation that the process interaction, combined with the participants' knowledge of each other, will enable individuals to influence others, even though the votes are equally weighted. More powerful members of the group can persuade others to vote as they do. We recognize that there are two kinds of power present in a group setting: the ability to vote which is equal among members and the ability to influence votes which is unequal among members.

3.4 Concealed or Distorted Preferences

Participants in a group decision process are not always willing to reveal their true preferences. In fact, they may wish to conceal the most important item on their agenda because explicitly stating it might lead to its defeat, e.g., if it becomes a focus for the opposition. Therefore, decision makers may support other proposals or present reasons that obscure their true thinking. They may even lend support to an extreme position in the hope that it will be defeated and a very different solution will be adopted. If participants are unwilling to state their true preferences, then the definition of issues will be incomplete and the analysis and priority setting in the AHP will be flawed.

One way of dealing with hidden agendas is to assemble a diverse decision-making group that includes a broad range of participants. In such a group, some participants may be able to uncover the hidden agendas of their "opponents." Another strategy is to design the rules of the decision-making sessions so that the final outcome or set of weights can only emerge from the set of stated issues.

Participants can also conceal preferences by intentionally distorting a position or opinion on an issue. Because the AHP uses comparisons to determine weights, participants can cast "extreme" votes in order to raise or lower the score produced by the geometric mean. Such votes would not reflect the actual assessment of the pairwise comparison but would represent an effort to distort the final score. This may happen when an individual perceives that he is in a minority position or that there are diametrically opposed opinions in the group.

When a group is evenly split between opposite positions, the group will easily anticipate the compromise value that results and will understand that an exaggeration of votes from one side can be countered by an exaggeration of votes on the opposite side. A group's reaction to a single extreme position is less predictable. If a group is convinced by an individual's argument that the position is sincere, then the outlying vote will be seen as legitimate. However, if the individual is perceived to be a manipulator or trouble-maker, then the vote could lead to retaliatory shifts in the majority consensus position. Thus, the danger of distortion may be greater in the group's reaction to one extreme voter than in the previous case. Often, disruptive judgments can be identified by their inconsistency with the remaining comparisons and the individual may be challenged to justify such extreme thinking.

A decision-making group may decide to police itself by appointing referees who can judge the motivation of extreme voters and rule on allowable values. It is important to specify the ground rules in advance so that arguments about the rules during the decision-making sessions can be avoided. (Arguments of this type tend to jeopardize the willingness of the group to engage in the actual process.) For example, consensus can be used for those judgments on which there is substantial agreement, while more controversial judgments may be synthesized from individual responses. The effects of intentional distortion on the outcomes of the process can also be diminished if, after the initial structuring of the problem, the entire group of decision makers is divided into several small groups. The smaller groups can make the pairwise comparisons simultaneously and separately and the final results can be obtained by averaging over the groups.

Finally, any group process is susceptible to breakdown when several participants refuse to continue. As mentioned before, setting ground rules and choosing referees would serve to give a group the ability to reprimand and remove disrupters. If a sizable number of participants are involved in disruptive behavior, it may be better to form several small groups in which these individuals are isolated in a single sub-group. Other groups can then proceed without interruption.

4. IMPLEMENTING THE RESULTS

After the final results have been generated, the decision-making group should evaluate the effort and cost of implementing the highest priority outcome. The group needs to determine whether it is likely that the participants and their constituencies will cooperate in the implementation phase of the effort. To be useful, the decision-making process must be acceptable to the participants and they must be willing to abide by the outcome. Finally, it is important for the group to view the AHP not as a tool for isolated, one-time applications, but rather as a process that has ongoing validity and usefulness to an organization. The AHP permits iterations and adaptations that can incorporate changing environmental factors. The process can help participants identify those variables that are subject to rapid change and even help the group attach probabilities to those changes.

5. REFERENCES

1. J. Aczel and T. Saaty, "Procedures for Synthesizing Ratio Judgements," **Journal of Mathematical Psychology, 27,** 93-102 (1983).

2. G. DeSanctis and R. Gallupe, "A Foundation for the Study of Group Decision Support Systems," **Management Science, 33,** 589-609 (1987).

3. P. Harker, "Alternative Modes of Questioning in the Analytic Hierarchy Process," **Mathematical Modelling, 9,** 353-360 (1987).

4. P. Harker, "Incomplete Pairwise Comparisons in the Analytic Hierarchy Process," **Mathematical Modelling, 9,** 837-848 (1987).

5. G. Huber and R. McDaniel, "The Decision-Making Paradigm of Organizational Design," **Management Science, 32,** 572-589 (1986).

AN ALTERNATE MEASURE OF CONSISTENCY

Bruce L. Golden and Qiwen Wang
College of Business and Management
University of Maryland
College Park, Maryland 20742

ABSTRACT

The AHP provides a decision maker with a way of examining the consistency of entries in a pairwise comparison matrix and the hierarchy as a whole through the consistency ratio measure. It has always seemed to us that this commonly-used measure could be improved upon. The purpose of this paper is to present an alternate measure of consistency and demonstrate how it might be applied. The contributions and limitations of the new measure are discussed.

1. MOTIVATION

The traditional eigenvector method for estimating weights in the AHP yields a way of measuring the consistency of a decision maker's entries in a pairwise comparison matrix. As discussed by Harker in this volume, the consistency index (C.I.) is defined by

$$C.I. = (\lambda_{max} - N)/(N - 1)$$

where λ_{max} is the largest eigenvalue of an N x N pairwise comparison matrix. Saaty [1] has shown that if a decision maker is perfectly consistent in specifying the entries, then $\lambda_{max} = N$ and C.I. = 0. If the decision maker is inconsistent, then $\lambda_{max} > N$ and Saaty [1] has proposed the following consistency ratio (C.R.) to measure the degree of inconsistency:

$$C.R. = C.I./R.I.$$

This measure is formed by taking the ratio of the C.I. for an N x N matrix filled in by a decision maker to an average C.I. value (known as the random index or R.I.) computed from 500 N x N positive reciprocal pairwise comparison matrices whose entries were randomly generated using the 1 to 9 scale. A value of C.R. under 0.10 is taken to indicate the decision maker has been sufficiently consistent in specifying entries for the matrix.

The consistency ratio and the 10% cut-off rule are reasonable measures but, at the same time, somewhat arbitrary. Several questions such as the following come to mind:

1. Does it make sense to compare matrix entries against purely random entries? The consistency ratio does this in the sense that it compares a numerator, C.I., with a denominator, R.I., which is an average over 500 matrices with purely random entries.
2. Why 10%?

3. Should the cut-off rule be a function of matrix size?

4. If a decision maker used another approach to estimate weights, such as the row geometric mean procedure (see Saaty [1]), how would he measure consistency?

We invite the reader to consider the comparison matrix below. Is it consistent?

$$\begin{bmatrix} 1 & 3 & 6 \\ 1/3 & 1 & 5 \\ 1/6 & 1/5 & 1 \end{bmatrix}$$

Most readers would answer negatively. After all, once the first row is filled in, only the entry in row 2 and column 3 remains to be specified. This entry deviates by 150% from perfect consistency. However, if we solve for the largest eigenvalue which turns out to be 3.094 and use the R.I. given by Saaty [1], we obtain C.R. = .08 < .10. One can generalize from this and similar examples that the 10% cut-off rule is too easy to satisfy for small matrices and too hard to satisfy for large matrices.

In this paper, we seek to develop and demonstrate a measure of consistency with the following four properties:

* The measure is easy to use;
* The measure can be used in conjunction with the traditional eigenvector method or the simpler row geometric mean method for estimating weights;
* The measure of consistency is a function of matrix size;
* The underlying probability distribution is intuitively appealing.

With respect to the last property, we attempt to construct a probabilistic model of how intelligent decision makers choose comparison matrix entries. We assume the intelligent decision maker can do much better than input purely random entries.

2. EXPERIMENTS WITH RANDOM MATRICES

In Saaty [1], the R.I. is reported for N = 3 to N = 15 where N is the dimension of the square comparison matrix. These numbers actually represent a composite of two different experiments--one performed at the Oak Ridge National Laboratory (ORNL) and the other performed at the Wharton School. More specifically, 500 random matrices were generated for N = 3 to N = 11 at the Wharton School; 100 random matrices were generated for N = 3 to N = 15 at the ORNL. The composite consists of the Wharton results plus the ORNL results for N = 12 to N = 15. These results and our more comprehensive experimental results are displayed in Table 1.

Table 1. Mean and Standard Deviation of R.I.

	R.I.			St. Dev.		Acceptance Probability
Source	Golden & Wang	Wharton	Oak Ridge	Golden & Wang	Oak Ridge	Golden & Wang
Sample Size	1000	500	100	1000	100	1000
N = 3	0.5799	0.58	0.382	0.7381	0.5165	0.225
4	0.8921	0.90	0.946	0.6299	0.6580	0.038
5	1.1159	1.12	1.220	0.5243	0.5280	0.0
6	1.2358	1.24	1.032	0.4120	0.4247	0.0
7	1.3322	1.32	1.468	0.3358	0.3478	0.0
8	1.3952	1.41	1.402	0.2895	0.2719	0.0
9	1.4537	1.45	1.350	0.2382	0.2190	0.0
10	1.4882	1.49	1.464	0.2131	0.1691	0.0
11	1.5117	1.51	1.576	0.1946	0.2161	0.0
12	1.5356		1.476	0.1663	0.5634	0.0
13	1.5571		1.564	0.1540	0.1750	0.0
14	1.5714		1.568	0.1360	0.1413	0.0
15	1.5831		1.586	0.1311	0.1457	0.0

In our experiment, we computed the average and standard deviation over 1000 random matrices for N = 3 to N = 15. Since our sample size is larger than in previous studies, we use our R.I. values throughout this paper. In addition to reporting the mean and standard deviation of C.I. over random matrices, Table 1 includes the acceptance probability as a function of matrix size. For example, for N = 3, 225 of the 1000 random matrices satisfied the 10% rule. The probability that C.I. is less than 10% of the R.I. exceeds .2 at N = 3, but drops to near zero at N > 4.

3. A NEW MEASURE OF CONSISTENCY

Consider the matrix of comparisons in the AHP

$$C = \begin{bmatrix} C_{11} & C_{12} & \cdots & C_{1N} \\ C_{21} & C_{22} & \cdots & C_{2N} \\ \vdots & \vdots & & \vdots \\ C_{N1} & C_{N2} & \cdots & C_{NN} \end{bmatrix}$$

where $C_{ii} = 1$ and $C_{ij} = 1/C_{ji}$ for $i,j = 1,2,\ldots,N$. The row geometric mean procedure works as follows. Let

$$g = \begin{bmatrix} g_1 \\ g_2 \\ \vdots \\ g_N \end{bmatrix} = \begin{bmatrix} \sqrt[N]{C_{11}\, C_{12} \cdots C_{1N}} \\ \sqrt[N]{C_{21}\, C_{22} \cdots C_{2N}} \\ \vdots \\ \sqrt[N]{C_{N1}\, C_{N2} \cdots C_{NN}} \end{bmatrix}$$

and normalize g such that

$$g^* = \begin{bmatrix} g_1^* \\ g_2^* \\ \vdots \\ g_N^* \end{bmatrix} = \begin{bmatrix} g_1/(g_1+g_2+\cdots+g_N) \\ g_2/(g_1+g_2+\cdots+g_N) \\ \vdots \\ g_N/(g_1+g_2+\cdots+g_N) \end{bmatrix}.$$

We also normalize each column vector of C so that $C^* = (C_1^*, C_2^*, \ldots, C_N^*)$ where

$$C_j^* = \begin{bmatrix} C_{1j}^* \\ C_{2j}^* \\ \vdots \\ C_{Nj}^* \end{bmatrix} = \begin{bmatrix} C_{1j}/(C_{1j}+C_{2j}+\cdots+C_{Nj}) \\ C_{2j}/(C_{1j}+C_{2j}+\cdots+C_{Nj}) \\ \vdots \\ C_{Nj}/(C_{1j}+C_{2j}+\cdots+C_{Nj}) \end{bmatrix} \quad \text{for } j=1, \ldots, N.$$

To measure consistency, we can use the average of absolute deviations of C_j^* from g^*. That is, let $G = 1/N \sum_i \sum_j | C_{ij}^* - g_i^* |$ where i and j range from 1 to N. If C is perfectly consistent, then $G = 0$. We remark that if a user prefers the traditional weight estimation approach, he can generate an eigenvector corresponding to the largest eigenvalue and normalize that vector such that its elements sum to unity. If we denote the resulting vector by

$$e^* = \begin{bmatrix} e_1^* \\ e_2^* \\ \vdots \\ e_N^* \end{bmatrix},$$

then the analogous consistency measure $E = 1/N \sum_i \sum_j | C_{ij}^* - e_i^* |$ can be used.

We tested the new measure of consistency, G, via a simulation experiment. A key goal was to construct a realistic probability distribution for G. In other words, if 1000 intelligent decision makers were filling in an N by N matrix, what would the frequency histogram look like? Our proposed probability model assumes that the user will enter a value in each row $r > 1$ that is at least close to the perfectly consistent entry implied by row 1. More specifically, each trial was performed in the following way:

1. Let $C_{ii} = 1$ for $i = 1, \ldots, N$. Let T be the empty set.
2. Randomly choose C_{1j} from $S = \{1/9, 1/8, 1/7, \ldots, 1/2, 1, 2, 3, \ldots, 9\}$ for $j = 2, 3, \ldots, N$.
3. For $i = 2,3,\ldots,N - 1$ and $j = i+1,\ldots,N$, do (a) through (f).

    a. For C_{ij}, let C_{1j}/C_{1i} represent perfect consistency.

    b. Choose the k largest numbers from S that are less than C_{1j}/C_{1i} and include these in the set T.

    c. Choose the k smallest numbers from S that are larger than C_{1j}/C_{1i} and include these in T also.

    d. If $C_{1j}/C_{1i} \in S$, then include this ratio in T also.

    e. If we cannot find k numbers in (b) or (c), include as many in T as possible.

    f. Randomly select a number from T. Let C_{ij} equal this number and let $C_{ji} = 1/C_{ij}$.

4. Compute G.

In our experiments, we decided that the most reasonable choice for k was 3. Essentially, the probability distribution described in steps 1 through 3 asks the following question: Given that the decision maker is trying to be consistent, will he succeed? Setting k to 3 indicates that the decision maker is really trying to be consistent. For less sophisticated decision makers, $k = 4$ or $k = 5$ may be more appropriate. With $k = 1$, or $k = 2$, it is difficult for the decision maker to be anything but consistent.

Our simulation experiment was performed for $k = 3$ and N varying from 3 to 11. The sample size was 1000 for each value of N. The frequency histograms of G are displayed in Figure 1. It is easily seen that G is approximately normally distributed for $N \geq 4$. We formally tested normality using the chi-square and Kolmogorov-Smirnov tests. The results in Table 2 indicate that the normal distribution provides a good fit, except for the case where $N = 3$.

Another detailed simulation experiment was performed to test the distribution of E. Although we do not present the results here, they were fully comparable to those of Figure 1 and Table 2. In other words, E is also approximately normally distributed for $N \geq 4$.

Table 2. Normality Test for G

	Chi-Square Test				K-S Test		
N	Statistic	Degrees of Freedom	Critical (95%)	Value (99%)	Statistic	Critical (95%)	Value (99%)
3	125.75	26	38.89	45.64	2.11	0.895	1.035
4	28.20	21	32.67	38.93	.98	0.895	1.035
5	15.46	24	36.42	42.98	.53	0.895	1.035
6	8.95	19	30.14	36.19	.53	0.895	1.035
7	17.12	20	31.41	37.57	.49	0.895	1.035
8	17.71	23	35.17	41.64	.44	0.895	1.035
9	13.23	20	31.41	37.57	.48	0.895	1.035
10	20.66	18	28.87	34.81	.60	0.895	1.035
11	25.00	22	33.92	40.29	.68	0.895	1.035

In a further experiment, we sought to compare g^* and e^* by computing the absolute deviation between the two vectors, given by

$$d = \sum_i |g_i^* - e_i^*|.$$

We then averaged over the sample of size 1000; the results are displayed in Table 3. They indicate that the average absolute deviation is less than 4% of the sum of elements of each vector.

Figure 1. Frequency Histograms of G

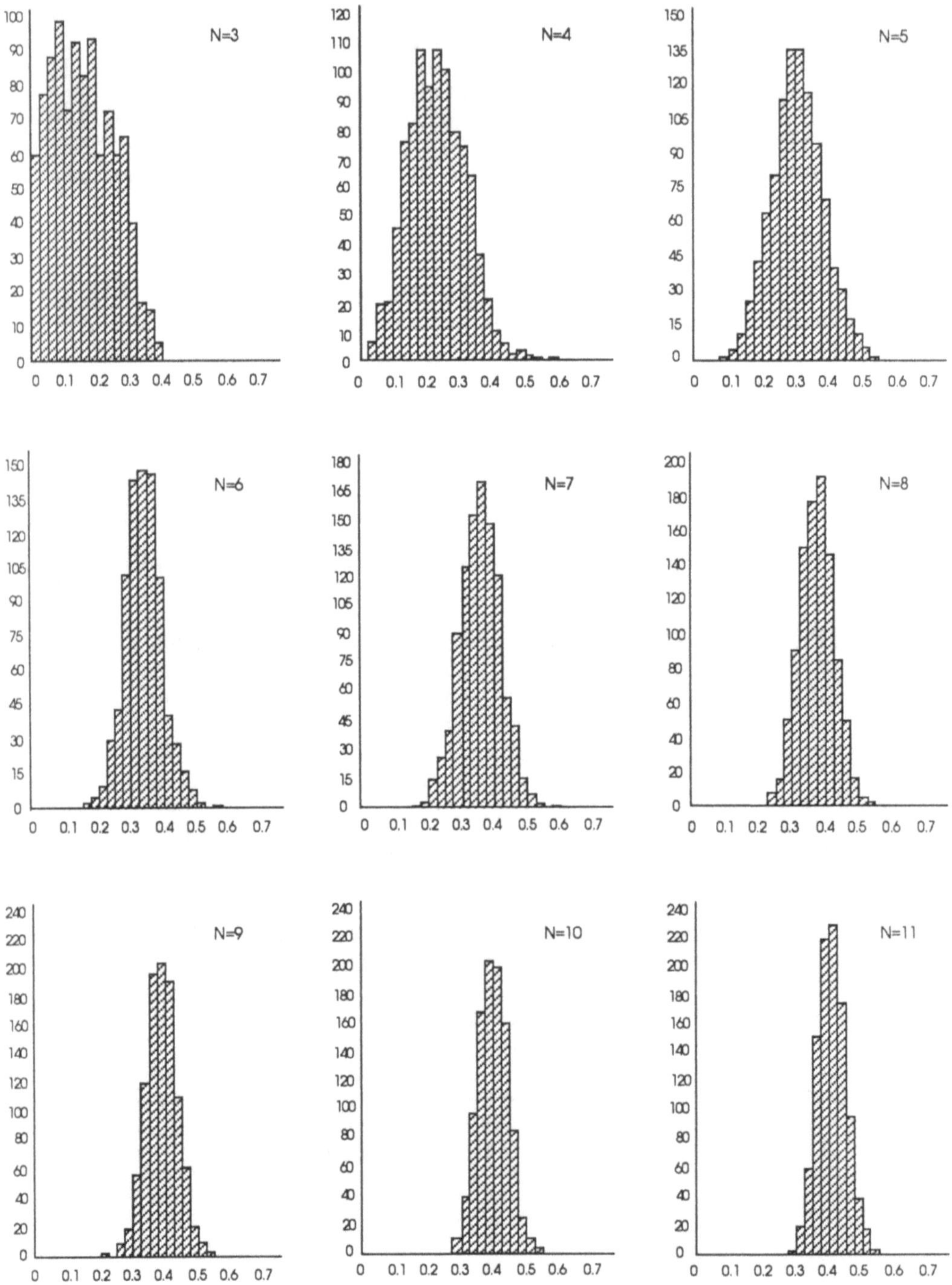

Table 3. Average Absolute Deviations Between g^* and e^*

N	3	4	5	6	7	8	9	10	11
Average Absolute Deviation	.000	.020	.028	.033	.036	.037	.038	.038	.038

In order to fully specify how the new consistency measure is to be used, we need to recommend a critical value. Based on a variety of computational experiments, we found that the 33rd percentile of the distribution of G works quite well. One way to think of this is as follows: The choice of k = 3 indicates that the decision maker is trying to be consistent. The null hypothesis is that he fails. If, for a matrix, the new consistency measure is to the left of the recommended critical value, then the null hypothesis is rejected.

In Table 4, the cut-off values derived from the 33rd percentile rule are provided for N varying from 3 to 11. Actually, three cut-off values are given for each value of N. P(33%) denotes the 33rd percentile found directly from the simulation experiment. Ns (33%) is an approximation to the 33rd percentile based on the simulation experiment. The simulation gives rise to an average and standard deviation (S.D.) for G. Using these and the normality assumption, Ns (33%) is easily computed. From the data, we were also able to use regression to obtain closed-form estimates for the average and standard deviation of G:

Average = 0.92781 + 0.3987298*Ln(N-1) - 0.2034568*(N-2)**.5 ;
Standard Deviation = 0.1327048 - 0.0304716*(N-2)**.5 .

In each case, R^2 exceeded .99. Using the above expressions and the normality assumption, Nr (33%) is again easily computed.

Table 4. Cut-Off Point for New Consistency Measure

	Simulation Results				Regression Results		
N	P(33%)	Average	S.D.	Ns(33%)	Average	S.D.	Nr(33%)
3	0.103	0.1642	0.09964	0.1204	0.1657	0.1022	0.1207
4	0.196	0.2446	0.09409	0.2032	0.2431	0.0896	0.2037
5	0.259	0.2945	0.08159	0.2586	0.2931	0.0799	0.2580
6	0.299	0.3303	0.07090	0.2991	0.3276	0.0718	0.2960
7	0.323	0.3496	0.06323	0.3218	0.3523	0.0646	0.3239
8	0.344	0.3685	0.05524	0.3442	0.3703	0.0581	0.3448
9	0.361	0.3815	0.04980	0.3596	0.3836	0.0521	0.3607
10	0.371	0.3989	0.04777	0.3729	0.3934	0.0465	0.3729
11	0.381	0.4026	0.04379	0.3833	0.4005	0.0413	0.3824

Table 4 reveals that P(33%), Ns (33%), and Nr (33%) are nearly identical, except for N = 3. This indicates that it is safe to use any one of these three cut-off values in making consistency decisions.

Next, we seek to contrast the new measure of consistency with the old (traditional) 10% cut-off rule. For each value of N, we counted the cases where the two measures agree and disagree. More specifically, we defined the following counters:

F(R,R) = frequency that old measure rejects consistency and so does new measure;

F(R,A) = frequency that old measure rejects consistency and new measure accepts;

F(A,R) = frequency that old measure accepts consistency and new measure rejects;

F(A,A) = frequency that old measure accepts consistency and so does new measure;

F(A,*) = frequency that old measure accepts consistency;

F(*,A) = frequency that new measure accepts consistency.

These counters are exhibited in Table 5. As before, the sample size is 1000 for each value of N.

Table 5. Comparing the Old and New Consistency Measures

N	F(R,R)	F(R,A)	F(A,R)	F(A,A)	F(A,*)	F(*,A)
3	415	0	218	367	585	367
4	492	38	152	318	470	356
5	531	65	143	261	404	326
6	575	104	97	224	321	328
7	606	135	73	186	259	324
8	625	159	47	169	216	328
9	649	203	26	122	148	325
10	646	227	14	113	127	340
11	654	270	4	72	76	342

From Table 5, we can make several important observations:

1. For N = 3 through 8, the old and new measures agree about 80% of the time;
2. For $N \geq 9$, the two measures agree less frequently;
3. The old measure is too easy for small N and too hard for large N;
4. The new measure behaves essentially the same way for all values of N. (This is, of course, due to our choice of k = 3 and 33%.)

4. EXAMPLES OF THE NEW MEASURE OF CONSISTENCY

It is now clear how one applies the new measure of consistency to a matrix, one at a time. To illustrate how the new measure of consistency is applied to a hierarchy, we work through two examples. The first hierarchy is taken from Saaty, Vargas, and Barzilay [2] and represents the Iran hostage rescue decision.

In performing their analysis, the authors assumed that moderate success was likely. The two options were--do and don't go ahead with the rescue mission. The five comparison matrices are reproduced in Table 6. We point out that the old and new measures of consistency are included in this table. For Matrix 2 through Matrix 5, the consistency ratios and G values are all equal to zero.

The overall consistency index is

$$(1*.10426 + .150*0 + .545*0 + .046*0 + .259*0)/$$
$$(1 + .150 + .545 + .046 + .259) = .05213$$

and the overall random index is

$$(1*.8921 + .150*0 + .545*0 + .046*0 + .259*0)/$$
$$(1 + .150 + .545 + .046 + .259) = .44605.$$

Therefore, the overall consistency ratio is

$$.05213/.44605 = .11687 > .10.$$

Table 6. Comparison Matrices for Example 1

Matrix 1
Relative Priorities of the Factors

	1	2	3	4	Priorities Old	Priorities New
1. Hostages' lives	1	1/3	5	1/3	.150	.154
2. Carter's political life	3	1	7	4	.545	.540
3. Military costs	1/5	1/7	1	1/6	.046	.047
4. U.S. prestige	3	1/4	6	1	.259	.259

Consistency Index = .10426
Consistency Ratio = .11688
G value = .31535

Matrix 2
Go/No-go Priorities for Hostages' Lives

	Go	No-Go	Priorities Old	Priorities New
Go	1	1	.5	.5
No-go	1	1	.5	.5

Matrix 3
Go/No-go Priorities for Carter's Political Life

	Go	No-go	Priorities Old	Priorities New
Go	1	3	.75	.75
No-go	1/3	1	.25	.25

Matrix 4
Go/No-go Priorities for Military Costs

	Go	No-go	Priorities Old	Priorities New
Go	1	1/7	.125	.125
No-go	7	1	.875	.875

Matrix 5
Go/No-go Priorities for U.S. Prestige

	Go	No-go	Priorities Old	Priorities New
Go	1	4	.8	.8
No-go	1/4	1	.2	.2

The hierarchy is, thus, regarded as inconsistent using the traditional measure of consistency.

The overall geometric mean measure of consistency is computed based on the overall G value and the overall dimension. The overall G value is

$$(1*.31535 + .154*0 + .540*0 + .047*0 + .259*0)/$$
$$(1 + .154 + .540 + .047 + .259) = .15768$$

and the overall dimension is

$$(1*4 + .154*2 + .540*2 + .047*2 + .259*2)/$$
$$(1 + .154 + .540 + .047 + .259) = 3.$$

Since .15768 > Ns (33%) = .1204, the hierarchy is viewed as inconsistent with respect to the new measure of consistency.

Both procedures for measuring consistency, therefore, declare this AHP model to be inconsistent. It is easy to see that Matrix 1 is the cause. Despite the fact that this matrix is clearly inconsistent, the C.R. value is just beyond the 10% cut-off value. The G value, on the other hand, is well beyond Ns (33%).

For the second example, we modify Example 1 in a minor way. Matrix 1 is used again in Example 2. Matrices 2 through 5 are of dimension 3 x 3 in Example 2 to take into account a third (hypothetical) option in each case--delay. Matrices 2 through 5 are shown in Table 7.

The overall consistency index is

$$(1*.10426 + .150*.02685 + .545*.00461 + .046*.01073$$
$$+ .259*.00915)/(1 + .150 + .545 + .046 + .259) = .05683$$

and the overall random index is

$$(1*.8921 + .150*.5799 + .545*.5799 + .046*.5799 + .259*.5799)/$$
$$(1 + .150 + .545 + .046 + .259) = .736.$$

Therefore, the overall consistency ratio is

$$.05683/.736 = .0772 < .10 \quad .$$

Here we have a very surprising result. In Example 1, Matrices 2, 3, 4, and 5 were perfectly consistent and the overall AHP model was inconsistent. In Example 2, Matrices 2, 3, 4, and 5 are no longer perfectly consistent and yet the overall consistency ratio does not deteriorate. In fact, the AHP model becomes consistent.

For Example 2, the overall G value is

$$(1*.31535 + .154*.15946 + .540*.06378 + .047*.06028$$
$$+.259*.08927)/(1 + .154 + .540 + .047 + .259) = .20015$$

and the overall dimension is

$$(1*4 + .154*3 + .540*3 + .047*3 + .259*3)/$$
$$(1 + .154 + .540 + .047 + .259) = 3.5 \ .$$

Table 7. Comparison Matrices for Example 2

Matrix 2
Go/Delay/No-go Priorities for Hostages' Lives

	Go	Delay	No-go	Priorities Old	Priorities New
Go	1	2	1	.413	.413
Delay	1/2	1	1	.260	.260
No-go	1	1	1	.327	.327

C.I. = .02685; C.R. = .04630; G value = .15946

Matrix 3
Go/Delay/No-go Priorities for Carter's Political Life

	Go	Delay	No-go	Priorities Old	Priorities New
Go	1	2	3	.540	.540
Delay	1/2	1	2	.297	.297
No-go	1/3	1/2	1	.163	.163

C.I. = .00461; C.R. = .00795; G value = .06378

Matrix 4
Go/Delay/No-go Priorities for Military Costs

	Go	Delay	No-go	Priorities Old	Priorities New
Go	1	2	1/7	.131	.131
Delay	1/2	1	1/9	.076	.076
No-go	7	9	1	.793	.793

C.I. = .01073; C.R. = .01850; G value = .06028

Matrix 5
Go/Delay/No-go Priorities for U.S. Prestige

	Go	Delay	No-go	Priorities Old	Priorities New
Go	1	2	4	.558	.558
Delay	1/2	1	3	.320	.320
No-go	1/4	1/3	1	.122	.122

C.I. = .00915; C.R. = .01578; G value = .08927

Since Ns(33%) for N = 3.5 is approximately

$$(.1204 + .2032)/2 = .1618 < .20015,$$

the AHP model is viewed as inconsistent with respect to the new measure of consistency.

5. CONCLUSIONS

We have presented an alternate measure of consistency that can be applied to a matrix and to an AHP model (hierarchy). Furthermore, we have used the measure to evaluate two illustrative AHP models. For these two examples, the new measure produces results that are intuitively appealing; the old method yields counter-intuitive results.

The new measure is indeed easy to use. It involves the computation of geometric mean vectors (or eigenvectors), some additional arithmetic, and a table look-up.

The underlying probability distribution, although somewhat arbitrary (e.g., in the choice of k = 3 and 33%), seems to reflect the behavior of serious decision makers more accurately than one in which entries are purely random. Naturally, a table similar to Table 4 could be constructed for k = 4 and 25%. The key point, however, is that the choice of k = 3 and 33% is very reasonable and it produces results that, in general, make sense.

In addition, we point out that, using the new measure of consistency, the critical value grows with matrix dimension, as one would expect.

A final objection to the old measure of consistency is the fact that the relatively large standard deviations shown in Table 1 for both N = 3 and N = 4 indicate how imprecise the random index values are for small N. The new measure does not rely upon these random index values.

6. REFERENCES

1. T.L. Saaty, **The Analytic Hierarchy Process**, McGraw-Hill, New York, New York (1980).

2. T. Saaty, L. Vargas, and A. Barzilay, "High-level Decisions: A Lesson From the Iran Hostage Rescue Operation," **Decision Sciences, 13**, 185-206 (1982).

A DECISION SUPPORT APPROACH FOR R&D PROJECT SELECTION

Matthew J. Liberatore
Department of Management
College of Commerce and Finance
Villanova University
Villanova, Pennsylvania 19085

ABSTRACT

Research and development (R&D) is often a wellspring of new ideas and concepts leading to the development of commercially viable products and processes. An organization's future market and financial positions may depend in large measure on the R&D project proposals which are selected. A variety of financial, market, technical, and manufacturing criteria may influence the selection decision. The importance of specific criteria varies by type of R&D activity, and the extent to which a particular project supports business objectives. This paper describes an approach for modeling the R&D project selection decision using the Analytic Hierarchy Process. The AHP represents an improvement over other well-known scoring approaches since the criteria weights or priorities established by the AHP are not based on arbitrary scales, but use a ratio scale for human judgments. The paper begins with a brief review of the R&D project selection literature, leading to a description of the desired characteristics for a decision support system for project selection. For a specific R&D strategy, namely, new product development, an AHP model is developed using an illustrative example. For situations requiring a large number of projects to be evaluated, the AHP model is expanded to include a series of performance ratings for each criterion. The performance ratings and weights for each criterion are transferred to a spreadsheet program which produces the final project rankings. The resulting project priorities or scores are included in an integer programming model to assist in the project funding decisions. The relationship between the integer programming approach and a form of benefit-cost analysis is discussed and illustrated. Two extensions of the AHP approach are then presented. The first addresses situations requiring the evaluation of a broader set of project selection criteria. The second illustrates how the AHP project selection model can be linked to the strategic planning process through an analysis of the mission, objectives, and strategies of the business. The paper concludes with a discussion of future areas of research.

1. INTRODUCTION

R&D project selection is an important resource allocation decision in many firms. During the budgeting and planning cycle, there are many project proposals vying for an organization's scarce funds, manpower, and facilities. The results of the project selection decision can impact the organization now and for many years to come. For example, a firm's investment in R&D can be substantial, often as high as 10% or more of sales in certain high-technology industries. Since R&D is often the lifeblood of new products and processes, an organization's future market and financial position may depend in large measure on the project proposals which are selected. As a result, methods and systems which support R&D project selection have commanded the interest and attention of many analysts and managers.

Operations research and management science (OR/MS) models and methods have long been developed and applied to the R&D project selection decision. A review of the R&D project selection literature reads like a litany of well-known OR/MS techniques: scoring models; linear, nonlinear, integer and goal programming; and multiattribute utility theory. However, there appears to be a gap between the development and implementation of these various techniques. New approaches are required which address qualitative as well as quantitative factors, consider the strategy of the organization, and utilize the expertise of the managers involved in the decision-making process. This paper describes one such decision support approach for R&D project selection, based on applications and extensions of the Analytic Hierarchy Process.

2. BACKGROUND AND LITERATURE REVIEW

Many models and methods for R&D project selection have been developed and reviewed in various surveys (e.g., see [2, 3, 4, 5, 8, 12, 13, 30]). Questions concerning the use of these models and methods have consistently appeared in the literature (e.g., see [4, 8, 31]). Many methods have seen limited implementation because of their inability to address the diversity of project types and measurement criteria within specific organizations.

An empirical study [19] on the use of OR/MS and other quantitative techniques for R&D project selection by "Fortune 500" firms found : 1) a heavy usage of financial analysis techniques, such as payback period and net present value; 2) minimal usage of mathematical programming models; 3) limited usage of budgeting systems based on benefit-cost tradeoffs; and 4) moderate usage of some form of checklist or scoring method for project screening. Finally, the study found that many R&D managers do not perceive that the available methods appreciably improve their decision making.

These findings lead to three related conclusions concerning the development of R&D project selection methods and systems (see also [13]):

1) Consider the characteristics of the organization performing the R&D -- the business strategies and goals which the R&D activity supports must be addressed. Other factors include the availability of data for measuring costs and benefits, and matching the type of R&D being conducted (e.g., product versus process research) with an appropriate set of criteria.

2) Measure social benefit-cost as well as economic factors-- both qualitative and quantitative criteria may require consideration. For example, in the development of a new technology, patent position may be an important, but difficult to quantify, criterion. In pharmaceutical or pesticide R&D, regulatory compliance may be critical. This criterion is qualitative in nature, and has economic and social consequences. As a result, scoring models such as those in [9] and [10] were developed to consider the diversity of project selection criteria. Several authors (e.g., [6, 22, 23, 24]) have identified different sets of factors that affect project selection, and have described methods for scoring and aggregating these to obtain a single evaluation measure.

3) Use methods which measure and aggregate multiple criteria -- simply basing project selection decisions on one or two measures of projected financial return is insufficient. Multicriteria methods can assist in project selection, and include multiattribute utility theory (MAUT) (e.g., [14, 15, 21, 28]) and goal programming (e.g., [16, 32]).

Finally, in developing methods and systems for R&D project selection, it is important to explicitly recognize and incorporate the knowledge and expertise of the R&D manager and supporting staff. Emphasis should be placed on adopting a decision support approach for project selection so that the manager can help structure the relationships between objectives, selection criteria, and project proposals. Techniques such as MAUT and goal programming can be incorporated into such decision support systems. However, a refocusing from normative decision models to decision support and knowledge-based systems is required. For example, a system called ISMAUT (Imprecisely Specified Multiattribute Utility Theory) [28] uses natural language statements to elicit the information required to develop the probabilities, values, and weights required in a linear additive utility function. In what follows, we will consider a decision support approach for R&D project selection based on applications and extensions of the Analytic Hierarchy Process. Earlier applications of the AHP for R&D project selection are reported in [17, 18].

3. AN ILLUSTRATIVE EXAMPLE

Consider a hypothetical firm, Novatech, Inc., which manufactures and sells a line of fertilizers. Novatech represents a composite of real-world firms that the author has worked with in the chemical industry. The project selection criteria and scoring approach presented below are typical of the planning processes observed in several of these firms. The suggested AHP approach has been used by one aerospace firm and is currently being evaluated by several process firms.

Our discussion begins with the Novatech business planning team which is working on the development of a five year strategic plan for the fertilizer Strategic Business Unit (SBU). The planning team consists of the marketing, finance, and manufacturing managers for the SBU, the R&D manager who serves as the liaison for the SBU, and corporate representatives from information systems and commercial development. The team has decided that the R&D Department should focus on the development of new products within this business segment. This represents a change in strategy from an emphasis on cost reduction for the current product line. New product ideas may include, for example, a fertilizer designed for a specific family of shrubs, or one which offers slow release of its nutrients under certain weather and soil conditions, and so on. A series of project proposals have been prepared, and the business planning team along with the R&D director must decide which of the projects are to be funded.

In the past, Novatech has used a scoring method for ranking projects, as shown in Figure 1. Seven criteria were selected, and weights assigned to each. Each project proposal was scored with respect to each criterion. A weighted average score was then com-

Figure 1. Novatech's Scoring Model

CRITERIA	Points* (1 TO 4)	X	WEIGHT	=	SCORE
Market Share			3		
Meeting Facility and Equipment Requirements			1		
Probability of Technical Success			3		
Development Cost			2		
Development Time			2		
Capital Investment			1		
Return on Investment			3		

PROJECT TITLE ________________ TOTAL SCORE ______________

OBJECTIVES ________________ SCORED BY: ______________

DATE________________ DEPARTMENT ______________

*Points: 1 = below average, 2 = average, 3 = above average, 4 = outstanding.

puted and used to rank the projects. However, the planning team has not been satisfied with the scoring method used for project selection for several reasons.

First, several team members were dissatisfied with the way in which the weights for each criterion were determined. In fact, the information systems manager argued that the criteria weights should depend on the particular business strategy that the project proposals support. Second, the rating levels for each criterion were felt to be somewhat arbitrary. For example, does an average rating for market share have the same meaning as an average rating for net present value? That is, the values of each rating level, namely, outstanding, above average, average, and below average, should be developed with a specific criterion in mind.

Finally, simply multiplying the rating score times the criterion weight and summing over criteria was less than satisfying. The ratings were ordinal numbers, so the final project score cannot be viewed as a cardinal value. Questions arose concerning how to translate the project scores into funding decisions. In the past, projects were funded in descending order of project total score until the budget was depleted. The financial manager asked if some type of benefit-cost analysis could be used to aid in the funding decision. For these reasons, an alternate approach for ranking and funding projects was investigated.

Novatech must begin the decision process by choosing an appropriate set of project selection criteria given the change in strategy from cost reduction to product development. As indicated earlier, the selection criteria should reflect the dimensions of the business strategy that the projects support (see also [17]). Previously, an important project selection factor under the current product improvement strategy was market share. Since the business strategy now concerns new product development, the business team expressed a preference of market growth over market share as their key marketing criterion. The team decided to market the new products after only one or two years of R&D, so net present value was selected over other risk-adjusted measures of financial return, such as those based on certainty equivalents (as developed in utility and decision theory and applied in investment analysis).

After further discussion, the team restricted their consideration to products which are new to the company but not necessarily to the marketplace. As a result, Novatech's capability to market the product was felt to be an important criterion. The previous product improvement strategy did not include this factor. The manufacturing representative indicated that the necessary facilities and resources for the potential new products are already available, so facilities was not chosen as a criterion. Because of increased environmental concerns for all lawn care and gardening products, compliance with all government regulations was added to the list of criteria. Finally, the team agreed that the probability of technical success, product development cost, and capital investment outlay continue to be important criteria for R&D project selection. The set of project selection criteria chosen for the new product development strategy is summarized in Figure 2.

Figure 2. Novatech's Project Selection Criteria Under the New Product Development Strategy

NPV - Net Present Value of Cash Flow Generated

MKT GWTH - Market Growth Rate

CAP MKT - Capability to Market New Product

REG COMP - Ability to Comply with Government Regulations

DEV COST - New Product Development Cost

TECH SUC - Probability of Technical Success

CAPITAL - Capital Investment Required

4. AN AHP/INTEGER PROGRAMMING APPROACH

4.1 Solution Approach

The proposed methodology uses the AHP to determine the project priorities and zero-one integer programming to assist in the funding decision. The problem objective is to maximize total priority over all funded projects, subject to a budgetary constraint and possibly other restrictions. This approach maximizes total project benefits subject to costs not exceeding the budget and yields results similar to those obtained from a modified benefit-cost analysis (discussed later in this chapter). The paramters and variables are defined as follows:

X_i = 1 if project i is funded
= 0, otherwise;

P_i = priority or weight of project i determined by the AHP;

C_i = cost of project i; and

B = total budget available for funding projects.

The basic formulation of this knapsack problem [27] is:

$$\text{Max} \sum_i P_i X_i \qquad (1)$$

subject to:

$$\sum_i C_i X_i \leq B, \qquad (2)$$

$$X_i = 0,1 \text{ for all } i. \qquad (3)$$

A variety of additional constraints can be included to ensure certain characteristics in the mix of total projects funded. For example, constraints on the minimum and maximum number of projects funded could be added easily. The project ranking and funding decisions can be accomplished using microcomputer-based software. Expert Choice [11] and Lotus 1-2-3 [20] can be used to determine the project priorities, while the binary integer programming problem can be solved with a commercial software package such as LINDO [29].

4.2 Project Priorities

The first stage of the decision-making process requires the determination of the project priorities or weights. The AHP developed by Saaty [25, 26] provides a methodology for structuring the hierarchical relationships between strategy, selection criteria, ratings levels, and projects. The judgments are entered as pairwise comparisons of items on a given level to each of the items on the next higher level of the hierarchy. The AHP can accommodate both subjective and uncertain information, and allows the application of experience, insight, and intuition in a logical way.

Figure 3. An AHP Model for Ranking R&D Projects

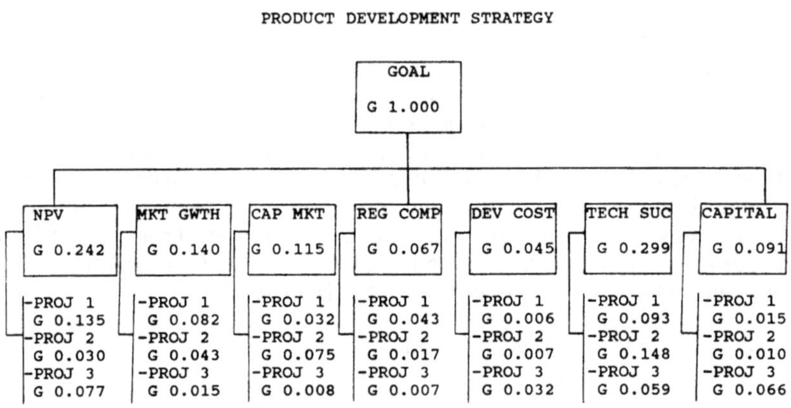

GOAL --- PRODUCT DEVELOPMENT STRATEGY
CAP MKT --- CAPABILITY TO MARKET
CAPITAL --- CAPITAL INVESTMENT
DEV COST --- DEVELOPMENT COST
MKT GWTH --- MARKET GROWTH
NPV --- NET PRESENT VALUE
PROJ 1 --- PROJECT 1 - BIOGEN
PROJ 2 --- PROJECT 2 - MISSYLINK
PROJ 3 --- PROJECT 3 - FERMENTATION SENSATION
REG COMP --- REGULATORY COMPLIANCE
TECH SUC --- PROBABILITY OF TECHNICAL SUCCESS

G --- GLOBAL PRIORITY: PRIORITY RELATIVE TO GOAL

OVERALL PROJECT PRIORITIES

PROJ 1	.405
PROJ 2	.329
PROJ 3	.265

The AHP hierarchy for R&D project ranking is shown in Figure 3. For purposes of illustration, we initially assume three R&D project proposals have been prepared. These are labelled BIOGEN, MISSYLINK, and FERMENTATION SENSATION. The details of these hypothetical proposals are omitted, since we wish to focus on the structuring and use of project selection methods.

The first step of the AHP analysis determines the importance of the R&D project selection criteria with respect to the goal of the R&D effort, namely new product development. The seven criteria were previously summarized in Figure 2. The required judgments can be obtained from the planning committee during a group decision-making process. These judgments are entered into Expert Choice as pairwise comparisons of the selection criteria relative to their importance in supporting the product development strategy. The planning team must be aware of the tendency of representatives from a given area such as marketing or finance to overestimate the importance of criteria from their respective functions. As a result, considerable discussion can be expected before a consensus is reached. "What if" analyses can be used to determine the impact of changes in the pairwise comparison data on the resulting priorities of the various criteria. Inconsistency of judgments can be measured and tracked using the features of Expert Choice. Fortunately, small deviations away from the implied ratio scale underlying the judgments leads to only small differences in the final priorities of the criteria and a generally stable solution.

The second stage requires pairwise comparisons of the three projects relative to each of the seven criteria. These judgments again require consensus among team members. If each team member rates the projects individually, one approach for combining the judgments is to compute their geometric mean, as suggested in Aczel and Saaty [1]. However, individual team members are responsible for the data necessary to support this decision-making process. For example, the financial manager is responsible for the net present value estimates and collaborates with the R&D manager in determining the capital investment levels required for each of the projects. The marketing manager develops the market growth estimates and works with the commercial development manager in determining project development costs and in evaluating the capability to market each of the potential new products. The probability of technical success for the projects is estimated by the R&D manager, who also evaluates regulatory compliance issues with the support of the corporate legal and environmental departments. The overall project priorities are obtained by summing across the criteria weights allocated to each of the projects, as summarized at the bottom of Figure 3.

In most organizations familiar to the author, more than three project proposals are evaluated as part of the annual budgeting and planning process. For example, in one industrial chemical firm familiar to the author, twenty or more project proposals were prepared for each of the larger business units. In such cases, pairwise comparing the projects on each selection criterion is generally tedious and time-consuming because of the large number

Figure 4. Examples of Ratings Description for Two Project Selection Criteria

PROBABILITY OF TECHNICAL SUCCESS

Rating	Probability Range
Outstanding	Over 70%
Above Average	Over 50% - Under 70%
Average	Over 30% - Under 50%
Below Average	Under 30%

CAPABILITY TO MARKET THE PRODUCT

Rating	Description
Outstanding	No problems are foreseen, since the product will be marketed mostly to known customers.
Above Average	Staff is generally familiar with the markets served by this product, although current products do not compete very much in these markets. No problems are foreseen.
Average	Staff is somewhat familiar with the markets served by this product, although the staff has little, if any, experience in these markets. No major problems are anticipated.
Below Average	Staff is completely unfamiliar with the markets served by this product, and the market patterns are sufficiently different to cause some startup problems.

---

of judgments required. To alleviate these difficulties, we suggest using a series of performance ratings for each selection criterion. We select ratings levels which are the same as in the scoring approach described previously: outstanding, above average, average, and below average. The business planning team can set ranges of numerical values or agree on detailed definitions to describe the four ratings levels for each criterion. Examples of ratings descriptions for the probability of technical success and capability to market the product are given in Figure 4.

However, unlike the scales used in the scoring method, pairwise comparisons between the four ratings levels are required for each of the criteria. These judgments are needed to maintain the ratio scale across the hierarchy. A sample judgment might be: With respect to the net present value criterion, how much more important is an outstanding rating than an average rating? These comparisons lead to weights for each of the four ratings levels associated with each criterion. These weights are then scaled by the weights of the criteria in achieving the goal of new product development so that a final or "global" weight for each rating level by criterion can be determined. The AHP model for R&D project selection using the ratings approach is given in Figure 5.

Figure 5. An AHP Model for R&D Project Selection Using Performance Ratings

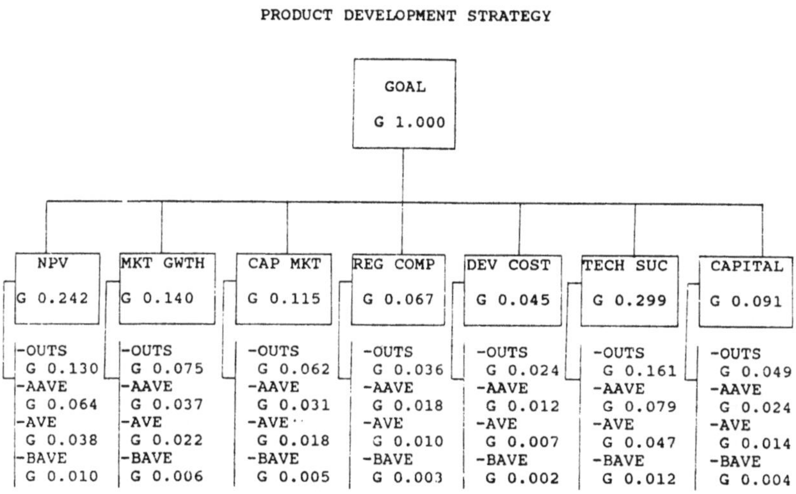

GOAL --- PRODUCT DEVELOPMENT STRATEGY
AAVE --- ABOVE AVERAGE
AVE --- AVERAGE
BAVE --- BELOW AVERAGE
CAP MKT --- CAPABILITY TO MARKET
CAPITAL --- CAPITAL INVESTMENT
DEV COST --- DEVELOPMENT COST
MKT GWTH --- MARKET GROWTH
NPV --- NET PRESENT VALUE
OUTS --- OUTSTANDING
REG COMP --- REGULATORY COMPLIANCE
TECH SUC --- PROBABILITY OF TECHNICAL SUCCESS

G --- GLOBAL PRIORITY: PRIORITY RELATIVE TO GOAL

*PROJECTS ARE NOT SHOWN SINCE THE RATING OF PROJECTS IS ACCOMPLISHED IN A SPREADSHEET (see TABLE 1) USING THE WEIGHTS DETERMINED IN THIS HIERARCHY

4.3 Project Rating Using a Spreadsheet

To complete the project rating process, the ratings levels and global weights for the project selection criteria are transferred to a spreadsheet in Lotus 1-2-3, as shown in Table 1. These data are located in the far left and far right columns of the spreadsheet, respectively. This spreadsheet is structured so that two columns are provided for each project. Here we assume that eight hypothetical projects must be ranked. A ratings level for a given criterion is selected by entering a "1" in the appropriate cell in the first of the two project columns. When the

Table 1. Project Rating Spreadsheet

CRITERIA	RATINGS LEVELS	PRODUCT DEVELOPMENT PROJECTS 1*	2	3	4	5	6	7	8	WEIGHTS**
NPV										0.242291
	OUTS	0	1 0.130	0	0	1 0.130	0	0	0	0.130236
	AAVE	1 0.064	0	0	0	0	0	1 0.064	0	0.064190
	AVE	0	0	1 0.037	1 0.037	0	1 0.037	0	0	0.037815
	BAVE	0	0	0	0	0	0	0	1 0.010	0.010048
MKT GWTH										0.140243
	OUTS	1 0.075	1 0.075	0	0	0	0	0	0	0.075384
	AAVE	0	0	0	1 0.037	1 0.037	0	1 0.037	0	0.037154
	AVE	0	0	1 0.021	0	0	1 0.021	0	1 0.021	0.021888
	BAVE	0	0	0	0	0	0	0	0	0.005816
CAP MKT										0.115294
	OUTS	0	0	0	0	0	0	0	0	0.061973
	AAVE	0	0	1 0.030	0	1 0.030	1 0.030	0	0	0.030545
	AVE	0	1 0.017	0	1 0.017	0	0	1 0.017	1 0.017	0.017994
	BAVE	1 0.004	0	0	0	0	0	0	0	0.004781
REG COMP										0.067059
	OUTS	0	1 0.036	0	0	0	0	0	1 0.036	0.036046
	AAVE	1 0.017	0	0	1 0.017	1 0.017	0	0	0	0.017766
	AVE	0	0	1 0.010	0	0	1 0.010	0	0	0.010466
	BAVE	0	0	0	0	0	0	1 0.002	0	0.002781
DEV COST										0.044329
	OUTS	0	0	0	0	0	0	0	0	0.024097
	AAVE	1 0.011	0	1 0.011	0	1 0.011	0	0	0	0.011676
	AVE	0	1 0.006	0	1 0.006	0	1 0.006	1 0.006	1 0.006	0.006996
	BAVE	0	0	0	0	0	0	0	0	0.001859
TECH SUC										0.299411
	OUTS	1 0.160	0	0	0	0	0	0	0	0.166940
	AAVE	0	1 0.079	1 0.079	1 0.079	0	0	0	1 0.079	0.079323
	AVE	0	0	0	0	1 0.046	1 0.046	1 0.046	0	0.046730
	BAVE	0	0	0	0	0	0	0	0	0.012417
CAPITAL										0.090869
	OUTS	0	1 0.048	0	0	0	0	0	0	0.048844
	AAVE	0	0	1 0.024	1 0.024	1 0.024	0	0	1 0.024	0.024074
	AVE	1 0.014	0	0	0	0	1 0.014	1 0.014	0	0.014182
	BAVE	0	0	0	0	0	0	0	0	0.003768
RAW SCORE		0.349	0.349	0.215	0.221	0.298	0.168	0.190	0.196	
RENORM SCORE		0.171	0.194	0.106	0.108	0.146	0.082	0.093	0.096	

* For each project, a 1 is entered under the first column to select a ratings level; the corresponding weight from the last column is entered in the second column when the spreadsheet is recalculated.

** From AHP model as shown in Figure 5.

spreadsheet is recalculated, the corresponding weight from the last column in the spreadsheet is moved into the cell next to the one in which the "1" was entered. The weights for the selected ratings are added for a total project score and renormalized to sum to one. This spreadsheet approach is similar in appearance to the scoring method, but it is important to remember that the ratings are based on the ratio scale underlying the AHP.

An alternative approach to <u>Lotus 1-2-3</u> is the RATINGS model provided with <u>Expert Choice</u>. However, <u>Lotus 1-2-3</u> was preferred by the author since it has: 1) the capability to interface directly with linear programming software such as <u>VINO</u> [7]; and 2) the computational functions needed to perform a modified benefit-cost analysis (described next). Another benefit of the suggested approach is that <u>Lotus 1-2-3</u> has become one of the most well-known and widely used analytical tools of business. A tie-in with <u>Lotus 1-2-3</u> should serve to facilitate the use of the proposed project selection method.

Table 2. R&D Project Funding Decisions

PROJECTS	RENORM. SCORE	PROJECT COST*	PRIORITY/ COST**	CUMUL. COST
5	0.146	300.00	4.867	300.00
1	0.171	420.00	4.071	720.00
4	0.108	270.00	4.000	990.00
2	0.194	500.00	3.880	1490.00
6	0.082	250.00	3.280	1740.00
3	0.106	340.00	3.118	2080.00
7	0.093	360.00	2.583	2440.00
8	0.096	400.00	2.400	2840.00
TOTAL	1.000	2840.00		

*In Thousands of Dollars
**Rescaled as a Number Between 0 and 10

4.4 Resource Allocation

The integer programming formulation given in equations (1)-(3) can be used to assist in the project funding decision. The renormalized project scores given in Table 1 are the weights or P_i's required in the objective function. The integer program for the illustrative example was run on LINDO at three funding levels: $1.490 million, $1.740 million, and $2.080 million. At the first budget level, projects 1,2,4 and 5 are funded, with project 6 and then project 3 added at the higher budget levels.

The funding decisions in each case are identical to those obtained by priority-cost analysis, as summarized in Table 2. Priority-cost analysis is a form of benefit-cost analysis, since the project priority, P_i, represents the sum total of the project's benefits as determined by the AHP. Priority-cost analysis is a "greedy heuristic" since it funds projects in non-increasing order of the priority-cost ratio P_i/C_i until the budget is depleted. However, differences between the two approaches can occur if the budget level is not exactly equal to a value of cumulative project costs given in non-increasing order of the priority-cost ratios. For example, if the budget equals $1.660 million, the greedy heuristic funds projects 1, 2, 4, and 5, while the optimal solution determined by integer programming funds projects 2, 3, 4, 5, and 6. In general, the greedy heuristic only provides a lower bound on the optimal solution value, although it often yields optimal or near-optimal solutions. Finally, we remark that there are "smart" heuristics for this problem which yield solutions which are very close to optimal (within 1 to 5%). The interested reader is referred to the work of Sahni [27].

5. AN EXTENSION OF THE BASIC APPROACH

In the previous example, the project selection criteria were specifically chosen to support a new product development strategy which was developed during the business planning process. In many firms, a single scoring model or checklist is used to rate all R&D projects, including product development, product improvement, process improvement, and exploratory research. Under these circumstances, the AHP approach presented in the previous section must be expanded to consider a more general set of criteria which are applicable to a whole range of R&D projects.

To evaluate a mix of R&D projects, it is useful to group the selection criteria into four general categories: technical, marketing, manufacturing, and financial (see also [18]). The technical criteria pertain to the R&D and engineering activities required to complete the project. This includes such subcriteria as probability of technical success, development cost, and ability to comply with government regulations, all of which were used in the previous example. Other subcriteria often considered in this category are development time, the availability of the required R&D and engineering resources, and patent status of the technology being developed or improved.

The marketing criterion addresses several factors previously considered, such as capability to market the product and market growth. Other marketing subcriteria that are often included in project selection methods are market share, size of potential or existing market, and customer acceptance. Manufacturing subcriteria which are frequently evaluated are capability of manufacturing the product, the ability of meeting the facility and equipment requirements, and manufacturing safety. There are numerous financial subcriteria that are usually considered, and these are often related to the firm's procedures for evaluating capital investments. Net present value and level of capital investment required are often used, but so are return on investment and unit cost of the product.

The AHP approach can easily accommodate this broader set of project selection characteristics. The resulting hierarchy has five levels: goal, criteria, subcriteria, ratings, and projects, as shown in Figure 6. R&D and business management must reach a consensus on the pairwise comparisons required at each level of the hierarchy. The last stage of the analysis requires that the projects be rated for each subcriteria using the spreadsheet approach described previously.

Figure 6. Expanded AHP Hierarchy for R&D Project Selection

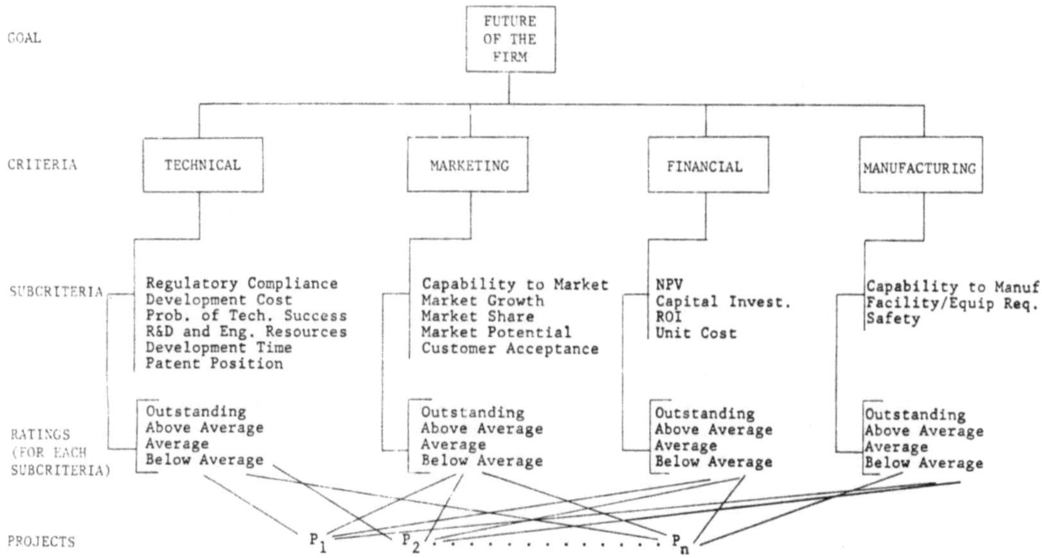

6. LINKING PROJECT SELECTION TO BUSINESS STRATEGY

Most organizations pursue several R&D strategies in parallel. For example, Novatech might decide to pursue product and process improvement strategies in addition to new product development. A series of project proposals might be put forth from R&D to support each of these three strategies. Using the AHP approach previously described, project priorities could be established for each strategy by developing three distinct project ratings spreadsheets. However, problems may arise in making the final project funding decisions, since tradeoffs may be required across the three sets of projects. This may occur, for example, if one budget is used to support all three strategies. Therefore, an analysis of the influence of each of these strategies in achieving Novatech's business objectives may be required.

The AHP approach can be used to address this problem. During the strategic planning process, Novatech's planning committee reviews the fertilizer SBU's mission, objectives, and strategies. Adjustments may be required, for example, if corporate management feels that the SBU's mission should be broadened to include a broader market base, or if higher levels of financial performance are desired. For example, the SBU's business mission might be to maintain its position as a leader in the North American home and garden fertilizer markets. Key business objectives over the next five years are to: 1) achieve a 10% return on investment; 2) reduce manufacturing costs by 15%; and 3) increase their share of the North American market by 5%. The planning committee can provide the necessary pairwise comparison information concerning the relative importance of these objectives in achieving the mission.

Next, the planning committee develops pairwise comparisons of the three R&D strategies for each of the business objectives. For example, the team might judge the process improvement strategy to be much more important than the product improvement and new product development strategies in achieving the cost reduction goal. However, the new product development strategy may be judged more important than the other two strategies in providing the market growth necessary to meet the market share objective. The pairwise comparisons needed to drive the AHP approach require the business planning team to refine these judgments in such a way as to represent the consensus of the group. Again, the geometric mean of the judgments might be used. The results of the analysis for this illustrative example are given in Figure 7.

So far, the AHP approach has addressed the priority of each R&D strategy in achieving the mission and objectives of the SBU. The linkage between R&D strategy and project selection is established as follows. The weights for each of the three R&D strategies shown in Figure 7 can be used to scale the scores of the supporting projects. For example, the renormalized project scores given in Table 1 would be multiplied by the weight of the new product development strategy (.280 as shown in Figure 7) to obtain the adjusted project scores. Similar computations would be completed for the other sets of project proposals supporting the product improvement and process improvement strategies. As previously discussed, integer programming or priority-cost analysis could be applied to assist in the final funding decisions across the three sets of project proposals. The application of the AHP to help weigh the various R&D strategies provides a direct linkage between a firm's business strategy and its project selection decisions.

7. SUMMARY AND CONCLUSIONS

R&D project selection is an important resource allocation decision in many firms. Although many quantitative techniques have been developed and/or applied to this problem, many have seen limited implementation because of their inability to address the diversity of project types and measurement criteria within specific organizations. Project selection methods should: 1) consider the characteristics of the organization performing the R&D; 2) measure social benefit-cost as well as economic factors; and 3) incorporate several different selection criteria. A suitable project selection method must also explicitly recognize and incorporate the knowledge and expertise of the R&D manager. A refocusing from a normative to a decision support approach is needed. The AHP is an appropriate modeling framework and can be used in conjunction with binary integer programming or priority-cost analysis to assist in the selection and funding decisions. The AHP is easy to use and can accommodate subjectivity and inconsistency in judgment. The AHP approach has been used by one of the largest U.S. aerospace firms to help reach a consensus on the ranking of R&D projects during their annual budgeting cycle.

Figure 7. An AHP Model for R&D Strategy

SBU MISSION: MAINTAIN POSITION IN N.A. HOME/GARDEN FERTILIZER MKT

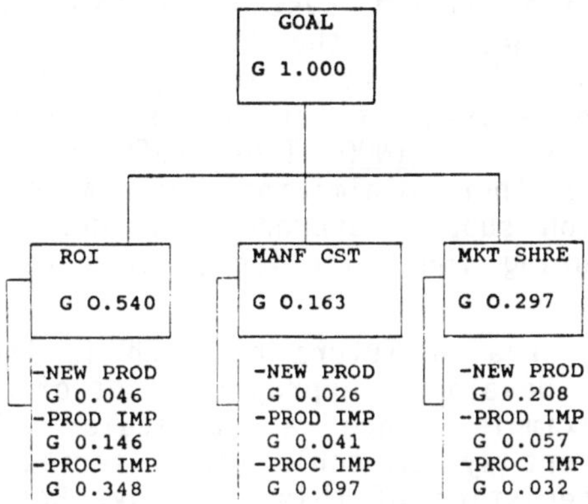

GOAL	---	SBU MISSION: MAINTAIN POSITION IN N.A. HOME/GARDEN FERTILIZER MKT
MANF CST	---	REDUCE MANUFACTURING COSTS BY 15%
MKT SHRE	---	INCREASE SHARE IN N.A. MARKET BY 5%
NEW PROD	---	R&D NEW PRODUCT DEVELOPMENT STRATEGY
PROC IMP	---	R&D PROCESS IMPROVEMENT STRATEGY
PROD IMP	---	R&D PRODUCT IMPROVEMENT STRATEGY
ROI	---	ACHIEVE 10% RETURN ON INVESTMENT
G	---	GLOBAL PRIORITY: PRIORITY RELATIVE TO GOAL

OVERALL R&D STRATEGY PRIORITIES

NEW PROD	.280
PROD IMP	.244
PROC IMP	.476

The suggested AHP approach is being evaluated by several process industry firms for similar application.

The combined AHP/integer programming approach is illustrated through an extended example. Situations requiring the ranking and evaluation of many projects are addressed through the development of project ratings for each selection criterion. A spreadsheet program directly linked to the AHP model assists the planning committee in developing the final project rankings. Integer programming is used to maximize total priority over all projects subject to a budgetary constraint. The relationship between the funding results obtained from integer programming and a simpler approach based on priority-cost analysis are explored.

Two extensions of the AHP approach are discussed. The first addresses situations requiring the evaluation of a broader set of project selection criteria. The second illustrates how the AHP project selection model can be linked to the mission, objectives, and strategies of the business. The latter requires a second AHP model to determine the weight of each of the various R&D strategies in achieving the mission of the business. These weights are used to scale the project scores developed for each class of R&D project proposals.

The combined AHP/integer programming method can form the basis of a decision support system for R&D project selection. Judgments are obtained from the R&D and business experts who develop the plans and strategies to support the business. This microcomputer-based system uses available software such as Expert Choice, Lotus 1-2-3, and LINDO. These software packages are easy to use and link together, minimizing programming and system problems. This decision support approach provides the needed flexibility in structuring the selection process and adapting it as needed.

An important area of future research is the use of expert system software to develop an appropriate set of project selection criteria. For example, once a particular R&D strategy is specified, an expert system could ask a series of questions to probe the characteristics of this strategy. Based on the user's responses, certain selection criteria would be chosen or additional questions asked. At the conclusion of the system-user dialogue, the system would provide the user with a ranked list of project selection criteria. Using the traceback feature available in most expert systems, the user could determine why a particular criterion was selected. This additional software support would further ease the development of the AHP/integer programming R&D project selection system.

8. REFERENCES

1. J. Aczel and T.L. Saaty, "Procedures for Synthesizing Ratio Judgments," **Journal of Mathematical Psychology, 27**, 93-102 (1983).

2. D. Augood, "A Review of R&D Evaluation Methods," **IEEE Transactions on Engineering Management, EM-20**, No. 4, 114-120 (1973).

3. N.R. Baker, "R&D Project Selection Models: An Assessment," **IEEE Transactions on Engineering Management, EM-21**, No. 4, 165-171 (1974).

4. N.R. Baker and W.H. Pound, "R&D Project Selection: Where We Stand," **IEEE Transactions on Engineering Management, EM-11**, No. 4, 124-134 (1964).

5. M.J. Cetron, J. Martino, and L. Roepke, "The Selection of R&D Program Content - Survey of Quantitative Methods," **IEEE Transactions on Engineering Management, EM-14**, No. 1, 4-13 (1967).

6. M.C. Cooper, "Evaluation Systems for Project Selection," **Research Management, 21**, 29-33 (1978).

7. K. Cunningham and L. Schrage, **Optimizing Lotus 123 with VINO**, The Scientific Press, Redwood City, California (1987).

8. B.V. Dean, **Evaluating, Selecting, and Controlling R&D Projects**, AMA Research Study 89, American Management Association (1968).

9. B.V. Dean and M. Nishry, "Scoring and Profitability Models for Evaluation and Selecting Engineering Projects," **Operations Research, 13**, No. 4, 550-569 (1965).

10. B.V. Dean and S.S. Sengupta, "Research Budgeting and Project Selection," **IRE Transactions on Engineering Management, EM-9**, 158-169 (1962).

11. E.H. Forman, T.L. Saaty, M.A. Selly, and R. Waldron, **Expert Choice**, Decision Support Software, McLean, Virginia (1983).

12. A.E. Gear, A.G. Lockett, and A.W. Pearson, "Analysis of Some Portfolio Selection Models for R&D," IEEE **Transactions on Engineering Management, EM-18**, No. 2, 66-67 (1971).

13. S.K. Gupta and L.R. Taube, "State of the Art Survey on Project Management," in **Project Management Methods and Studies**, B.V. Dean, editor, North-Holland, Amsterdam, The Netherlands (1985).

14. D.L. Keefer, "Allocation Planning for R&D with Uncertainty and Multiple Objectives," **IEEE Transactions Engineering Management, EM-25**, No. 1, 8-14 (1978).

15. D.L. Keefer and C.W. Kirkwood, "A Multiobjective Decision Analysis: Budget Planning for Product Engineering," **Journal of the Operational Research Society, 29**, No. 5, 435-442 (1978).

16. A.J. Keown, B.W. Taylor, and C.P. Duncan, "Allocation of Research and Development Funds: A Zero-One Goal Programming Approach," **Omega, 7**, 345-351 (1979).

17. M.J. Liberatore, "An Extension of the Analytic Hierarchy Process for Industrial R&D Project Selection and Resource Allocation," IEEE **Transactions on Engineering Management, EM-34**, No. 1, 12-18 (1987).

18. M.J. Liberatore, "R&D Project Selection," Telematics and **Informatics, 3**, No. 4, 289-300 (1987).

19. M.J. Liberatore and G.J. Titus, "The Practice of Management Science in R&D Project Management," **Management Science, 29**, No. 8, 962-974 (1983).

20. **Lotus 1-2-3** Release 2, Lotus Development Corp., Cambridge, Massachusetts (1985).

21. G.R. Madey and B.V. Dean, "Strategic Planning for Investment in R&D Using Decision Analysis and Mathematical Programming," **IEEE Transactions on Engineering Management, EM-32**, No. 2, 84-90 (1985)

22. D.B. Merrifield, "Selecting Projects for Commercial Success," **Research Management, 24**, 13-18 (1981).

23. A. Paolini, Jr. and M.A. Glaser, "Project Selection that Pick Winners," **Research Management, 20**, 26-29 (1977).

24. L.P. Plebani, Jr. and H.K. Jain, "Evaluating Research Proposals with Group Techniques," **Research Management, 24**, 34-38 (1981).

25. T.L. Saaty, **The Analytic Hierarchy Process**, McGraw-Hill, New York, New York (1980).

26. T.L. Saaty, **Decision Making for Leaders**, Lifetime Learning Publications, Belmont, California (1982).

27. S. Sahni, "Approximate Algorithms for the 0/1 Knapsack Problem," **Journal of the Association of Computing Machinery, 22**, No. 1, 115-124 (1975).

28. W.T. Scherer, B.S. Stewart, E.A. Sykes, and C.C. White III, "A New Interpretation of Alternative Pairwise Comparisons for a Generalization of SMART," **IEEE Transactions on Systems, Man and Cybernetics**, SMC **17**, No. 4, 666-670 (1987).

29. L. Schrage, **Linear, Integer and Quadratic Programming with Lindo**, 3rd ed., The Scientific Press, Redwood City, California (1986).

30. W.E. Souder, "Comparative Analysis of R&D Investment Models," **AIIE** Transactions, **4, No. 1, 57-64 (1972).**

31. W.E. Souder, "Analytical Effectiveness of Mathematical Models for Project Selection," **Management Science, 19**, No. 8, 907-923 (1973).

32. B.W. Taylor, L.J. Moore, and E.R. Clayton, "R&D Project Selection and Manpower Allocation with Integer Nonlinear Goal Programming," **Management Science, 28**, No. 10, 1149-1158 (1982).

PROJECT SELECTION BY AN INTEGRATED DECISION AID

Jukka Ruusunen and Raimo P. Hamalainen
Systems Analysis Laboratory
Helsinki University of Technology
Otakaari 1, SF-02150 Espoo, Finland

ABSTRACT

The use of the Analytic Hierarchy Process as part of a decision aid for research and development (R&D) project selection in a large Finnish company is described. The company's most important operational sector is the oil industry and a comprehensive petrochemical industry has been built as an extension to oil refining. Selection of R&D projects is a decision problem of vital importance in the company's long-range strategy. To evaluate one project proposal at a time, a subjective measurement scale is constructed for each lowest level criterion. The importance of the selection criteria as well as the measurement scales are assessed by the AHP or direct rating. The overall preference model is implemented in a decision aid by integrating different software modules. The preference models of the individual managers reside in a database which is managed by a database program. These models can be updated by a general purpose decision analysis program. The man-machine interface of the system is implemented by an expert system shell.

1. INTRODUCTION

The selection of R&D projects is a decision problem of vital importance in the long-range strategy of many industrial companies. The problem is to allocate the company's scarce resources over a set of project proposals in a situation where total resource requirements exceed those available. In today's rapidly changing markets, the R&D spendings are often a sizable investment with great uncertainty and risk. Besides risk, a characteristic feature of R&D project selection is the multiplicity of the relevant criteria ranging from the product's profitability and sales volume to technical risks in the manufacturing process.

Many companies use financial measures, e.g., payback period, return on investment, or cost/benefit analysis, to evaluate the benefits of a project. Scoring models are also used, in which one assumes that the project proposals can be evaluated in terms of a small number of decision criteria. Possible criteria are cost, manpower availability, and probability of technical success. The project in question is evaluated against each criterion and the resulting vector of scores is used to analyze the strengths and weaknesses of the project. An overall benefit measure is obtained from the scores usually by addition or multiplication. The complicating factors that are present in the problem can also be approached by structured techniques such as the AHP [11] and multiattribute utility theory techniques including value or utility measurement [12]. In the following discussion, the general term decision analysis is used for these approaches. The basic idea in decision analysis is to use problem decomposition and explicit value or preference trade-offs to improve the understanding of complex problems.

In spite of the fact that the potential usefulness of decision analysis in R&D project selection has been shown in the literature (see, e.g., [2,8]), there are very few reports about the practical implementation of the methods in decision support systems for R&D management. In fact, relatively little is known about the true usefulness of these systems in the R&D environment. Some of the pioneering works emphasized the choice of the methodology. However, today the R&D manager is certainly not interested in an academic battle between rival decision analysis methods. The interesting goal is not to show that a particular method can be successfully used in R&D evaluation, but to develop a system that really gains the acceptance of the users and can be easily maintained and updated even by the end users themselves.

In this paper, we will describe the use of decision analysis in R&D project selection in Neste Corporation, a large Finnish company that operates in the petrochemical industry. The company is the country's largest industrial enterprise with respect to turnover; the annual turnover is 6 billion U.S. dollars [9]. Crude oil imports, oil refining, and the marketing of oil constitute the company's most important operational sector. A comprehensive petrochemical industry has been built up as a natural extension to oil refining. The petrochemical plants produce ethylene, butadiene, propylene, benzene, phenol, and acetone. The company also produces polystyrene and industrial chemicals and has holdings in plastics companies.

The paper is organized as follows. The R&D activity of the Finnish company is described in Section 2. Section 3 describes the development of the decision aid from the structuring phase to the computer implementation. Experiences from working with the decision makers are reported in Section 4. However, due to confidentiality, we cannot present any quantitative data from our analysis.

2. R&D PROJECT SELECTION IN A FINNISH PETROCHEMICAL COMPANY

The company has a technology group for research and development. Together with development groups in the company's other divisions, the technology group devises new products and processes for use by the company, as well as catalysts and the methods for producing them. In addition to the development of new products, the main task of research is to back up the corporation's business units. This includes improving present products and production processes, both technically and economically, providing analysis and other special services, and the technical consulting required for marketing.

The corporation's R&D spendings are 80 million U.S. dollars [9]. The research activities often take place in totally new technological areas. The proportion of experimental development at the pilot stage has increased rapidly during the last few years. R&D has focused on the manufacture and product technology of polyolefins. New types of polyethylene catalysts have been manufactured and development of a new polymerization technique for polyethylenes has been started. Other examples of the research areas are conductive polymers, synthetic lubricants, and cellulose carbamate.

The company has a top level management group for R&D decisions. The members are managers of the company's various technological branches. The group's evaluations of projects have traditionally been based more on the personal insight and experience of the managers than on detailed quantitative studies. Still, the present analysis arose from the company's interest in learning about new modeling approaches to support group decisions.

The projects are classified into two groups: development projects supporting business ventures and research and development projects that take place in new technological areas. Examples of the first group are projects that raise the capacity of new plants, e.g., development of a new process which improves production of polyalphaolefins for the manufacture of synthetic lubricants. Projects that belong to the second group include research focused on polymers. The project groups have their own budgets and they are financed from different sources. There are flexible overall targets for the shares of individual project groups of the total R&D budget.

The company had previously experimented working with decision trees, where the problem is modeled as a sequence of risky choices. In spite of the difficulties found in the assessment of the probabilities in a decision tree, the managers felt that the method was potentially useful in decision aiding as far as short-term development projects are concerned. However, in the case of long-term strategic R&D projects which aim at totally new high-technology products, the decision tree approach did not correspond to management's way of problem solving.

3. SOLUTION METHODOLOGY

3.1 From an Expert System to Decision Aiding

Initially, the cooperative project described here arose from the company's interest in learning about the new possibilities that the emerging artificial intelligence technology could offer to strategic R&D decision making. Thus, a natural starting point was an attempt to develop a rule-based expert system for R&D project selection. We studied the possibility of defining rules and instructions about the factors that have to be considered in project evaluation on the basis of past experience. This corresponds to a computer implementation of checklists that are used in aiding the project selection process (see, e.g., [3]). However, even though checklists are easy for decision makers to understand, preferential differences in the importance of the selection criteria cannot be easily included in this kind of a simple approach. The available expert system shells were unsatisfactory because they did not help in the management of preferential knowledge. However, it turned out that their rule-based approach (see, e.g., [1]) is very useful in the implementation of the problem specific man-machine interface. What was still needed was the preference model that describes the importance of the selection criteria as well as the measurement scales for the criteria.

3.2 Problem Structuring

To structure the analysis, the R&D project selection problem is presented as a hierarchy of criteria, which consists of the relevant factors in the R&D project selection. The idea is to break down the original complex decision problem into its parts and thus allow the decision maker to focus on the different factors separately. Each level of the hierarchy consists of mutually independent criteria which are roughly of the same magnitude or importance. For example, such generally stated criteria as marketing and financing can be included at the same level of the hierarchy. These criteria are then broken down into lower level operational criteria. For example, marketing consists of the following criteria: the performance of the marketing organization, the commercial success of the product, and the effect of the product on existing products.

Depending on the techniques used to analyze the preferences in the hierarchical model, different terms for the approach are used: an objectives hierarchy [7], a hierarchy [11], or a value tree [12]. The names objectives hierarchy and value tree are used in the context of multiattribute utility theory and they have approximately the same meaning. In AHP's terminology, a hierarchy consists of the goal under which there are criteria, subcriteria, and alternatives [11]. Saaty [11] explains in detail how decision makers proceed through the structuring phase. The terminology used in the multiattribute utility theory approach is somewhat different. In general, objectives in an objectives hierarchy mean the same as criteria and subcriteria in the AHP. Similarly, the AHP's lowest level subcriteria correspond to attributes in the multiattribute utility theory formulation. In the multiattribute utility theory approach, the hierarchical decomposition of the problem is often only a structuring phase and preference tradeoffs are only evaluated at the lowest level. In the AHP, the structuring and ranking processes go more closely together. Saaty [11] has also proposed the generalization of the hierarchical model into a network of criteria and alternatives. The practical usefulness of this approach is just beginning to be studied [4,5,6]. This generalization of the AHP was not used in the present project.

The following criteria for examining the set of objectives and attributes have been proposed by Keeney and Raiffa [7]: *complete*, so that all relevant aspects of the problem are covered;

operational, so that the lowest level attributes are meaningful and assessable; *decomposable*, so that the problem can be broken down into parts; *nonredundant*, so that different attributes do not mean the same thing; and *minimal*, so that the number of attributes is as small as possible.

In the multiattribute utility theory approach, the measurement scales are constructed for each of the lowest level criteria. Individual decision alternatives are then rated by evaluating their performance using these existing measurement scales. The AHP is different from this approach in the sense that the alternatives to be rated form the lowest level of the hierarchy. Pairwise comparisons of the alternatives with respect to the lowest level criteria thus generate a single priority point. The comparisons are thus relative [10]. In the present problem, the R&D management often has to evaluate one project proposal at a time. Consequently, the representation that uses measurement scales for the lowest level criteria was chosen. The AHP was used to assess the importance of the criteria and also the measurement scales.

An absolute measurement scale extension to the basic AHP was recently proposed by Saaty [10]. Saaty uses absolute measurements to rate the alternatives in terms of rating levels of the lowest level criteria. The ratings or intensities of the criteria that he proposes to use are excellent, very good, good, average, below average, poor, and very poor. For each lowest level criterion, pairwise comparisons are performed on the ratings. An alternative is then evaluated by identifying the relevant rating for each lowest level criterion that best describes the alternative. A ratio scale score of the alternative is finally produced by adding the weighted priorities of the relevant ratings, one under each criterion corresponding to the alternative.

The approach that was used in the present project is more general than the above absolute measurement technique. In a more general setting, the intensities that describe the criteria are not the same for all the lowest level criteria, but are separately defined for each criterion. For example, if a criterion can be measured in monetary units, the rating levels should also be expressed in monetary units. Even when a criterion is measured on a subjective scale, the rating levels should be made as clear as possible to the evaluators. To avoid misunderstandings, this is particularly important in a group setting, which was the case in our project.

There is a technical problem with current AHP software that is due to this extension. The possible ratings of the criteria are included in the lowest level of the hierarchy. Consequently, when each criterion has its own intensities, the number of elements in the lowest level of the AHP hierarchy quickly becomes very large. In our case, it was impossible to manage the large number of elements with existing AHP software. This was one of the reasons why the man-machine interface of the decision aid was implemented using an expert system shell. Another difference between our approach and the AHP's absolute measurement is the scale for scoring the alternatives. As in the AHP, we performed pairwise comparisons on the rating levels for the lowest level criteria. The ratio was then transformed into a value scale between zero and one. The worst rating level is given the priority value zero and the best rating level is given the priority value one. Direct rating of the intensities on this scale was also possible. Using the terminology of multiattribute utility theory, this results in single-criterion value functions for all the lowest level criteria.

In the selection of the relevant criteria that affect the evaluation of R&D projects, the emphasis was on the long-range effects. Information was gathered from the R&D literature and from an interview with a contact person in the R&D management group. The criteria that we obtained are very similar to the general evaluation criteria described in the R&D project selection literature [3,8]. The resulting hierarchy reflects the three major factors in this decision problem:

1. ATTRACTIVENESS from the point of view of markets and the strengthening of the company's strategic position,

2. SUITABILITY with respect to the company's policy, marketing, financing, and manufacturing, and

3. LIKELIHOOD OF TECHNICAL SUCCESS from the point of view of R&D and production.

The *attractiveness* property is related to the product's expected future effects on the company's strategic position and on the product's future markets. Subcriteria are as follows:

1.1 Strengthening of strategic position
- strengthening of strategy
- renewal of image
- risks and threats
- technological significance

1.2 Marketing
- product's competitive environment
- sales volume
- market trend and growth.

The *suitability* property describes the product's suitability for the existing organization from the point of view of company policy, marketing, financing, and manufacturing. Subcriteria are as follows:

2.1 Company policy
- strategy
- image
- support from leadership
- interest groups

2.2 Marketing
- marketing organization
- commercial success
- effect on existing products

2.3 Financing
- capital investment
- availability of money
- profitability

2.4 Manufacturing
- type of production
- personnel
- equipment
- raw material.

The third main criterion is the *likelihood of technical success* of the project from the point of view of both R&D and manufacturing. Subcriteria are as follows:

3.1 Manufacturing
- available skills
- safety
- new equipment and processes

3.2 Research and development
- compatibility with the R&D strategy
- likelihood of success
- cost and time
- patent status.

The three main criteria with the lower level elements are shown in Figure 1.

3.3 Single-criterion Evaluations and Weighting

3.3.1 Single-criterion Evaluations

Because of the hierarchical decomposition, the decision maker can focus on the different lowest level criteria separately. Single-criterion evaluations are constructed corresponding to each of the lowest level criteria. This requires the construction of well-defined quantitative or qualitative measurement scales for these criteria. This scale is then converted into a value scale between zero and one by means of judgments about the relative desirability of the possible levels of the criteria. In our case, most of the criteria have a qualitative measurement scale. All of the scales are discrete.

In general, a value scale can be assessed using different decision analysis techniques. With the *numerical methods*, the decision maker directly assesses the strengths of preferences on a numerical scale. The AHP can be used as a numerical estimation method [11]. The value scale is then assessed by pairwise comparisons of the desirability of the possible levels of the lowest level criteria. Other methods that belong to this class

Figure 1. The Hierarchy of Criteria

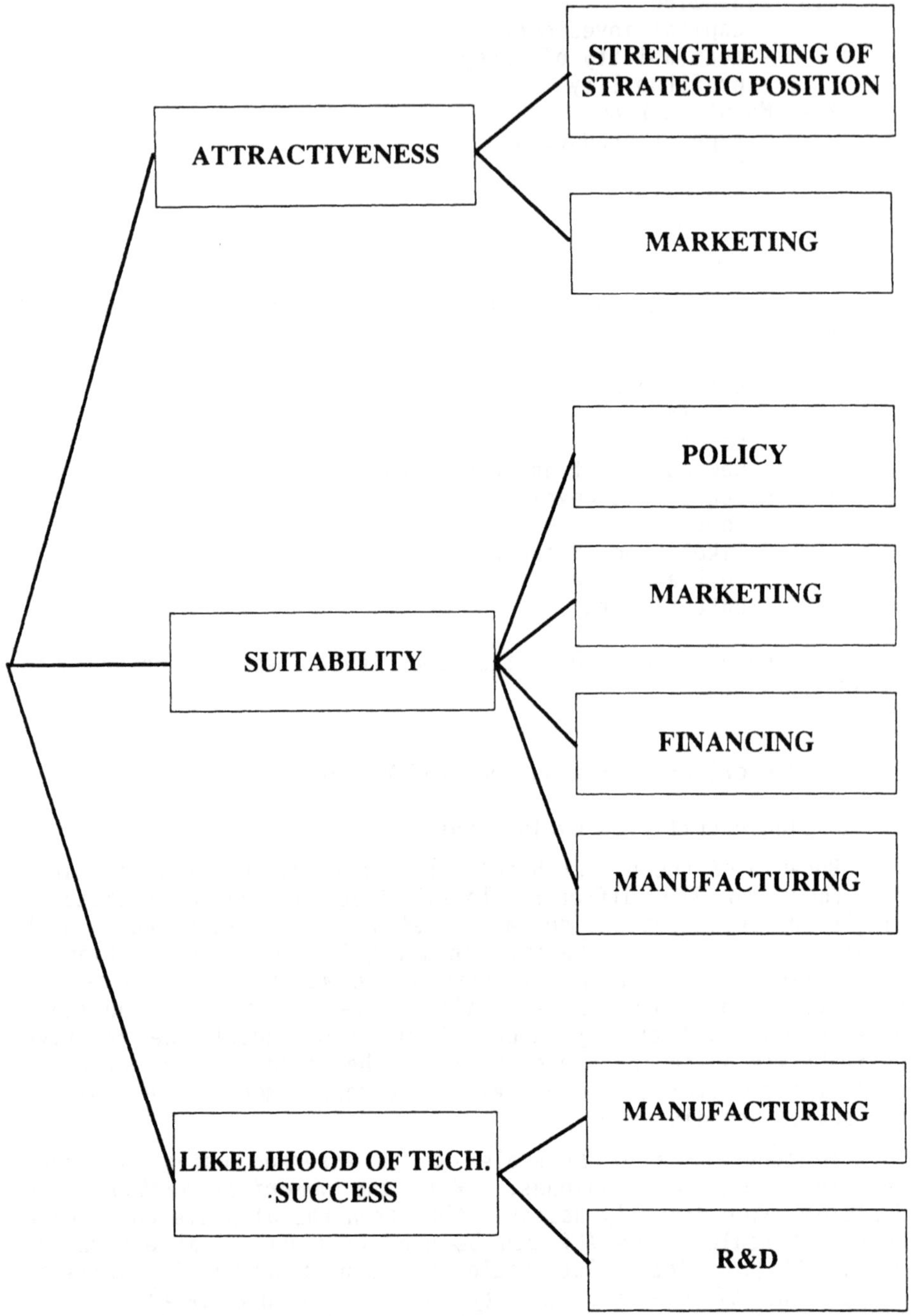

are direct rating, category estimation, ratio estimation, and curve drawing [12].

Another class of methods are the *indifference methods*, which are based on indifferences between pairs of evaluation objects. Methods in this class are the difference standard sequence method and bisection [7,12]. For example, in the bisection method the decision maker specifies midvalue points between two levels of a criterion such that an increase of the criterion from the lower level to the midvalue is preferentially equivalent to an increase from the midvalue to the higher level. Consider, for example, the profitability criterion which is measured in terms of the return on investment (ROI). Assume that the range of the criterion is from 10% to 30%. The subjective value scale corresponding to this criterion is given the value 0 at 10% and the value 1 at 30%. The decision maker first specifies the value of ROI, say x, between 10% and 30% such that the decision maker is indifferent between an increase in ROI from 10% to x and from x to 30%. The value scale is then assigned the value 0.5 at x. This procedure can then be continued with the intervals [10%,x] and [x,30%]. Further subdivision of the measurement scale is used to refine the value scale.

3.3.2 Weighting

To combine the single-criterion evaluations, tradeoffs among the criteria have to be determined. This can be done by different methods, which vary in the degree of complexity. In the direct methods, the decision maker usually determines the tradeoffs on the basis of the importance of the criteria relative to each other. The AHP is a direct method. In the AHP, the tradeoffs are evaluated by pairwise comparisons of the criteria. Another example is a ratio method, where the decision maker evaluates how much more important the criteria are relative to the least important one. This method is used in the multiattribute technique called SMART (see, e.g., [12]). In SMART, the single-criteria evaluations are based on direct rating.

3.4 Problem Solving Procedure

In general, the group's problem solving procedure includes the following steps.

i) The decision hierarchy is constructed such that it is accepted by the decision makers as representing all the relevant factors of R&D project selection. The ranges of measurement scales for the lowest level criteria are selected.

ii) Each decision maker constructs the single-criterion value scales for the lowest level criteria and weights the criteria. Lowest level weights and the corresponding single-criterion values are aggregated with the additive model

$$v(\underline{x}) = \sum_{i=1}^{n} w_i v_i(x_i),$$

where x_i is the measurement of the project on lowest level criterion i, $1 \leq i \leq n$, $\underline{x} = (x_1,\ldots,x_n)$, v_i is the corresponding single-criterion value scale, w_i is the corresponding weight, and $v(\underline{x})$ is the overall value of $\underline{x}$.

iii) Individual differences in the evaluations are discussed. The individual evaluations are then combined to obtain the group preference model.

iv) Project proposals are evaluated by using the group preference model.

Steps (i)-(iii) of the procedure are repeated when needed in order to update the group preference model, which is then used in step (iv) to evaluate new product proposals.

3.5 Implementation of the Decision Aid

Systems analysts have traditionally written their programs in general purpose programming languages such as Fortran, Pascal, or C. The advantage of this approach is generality, but writing and debugging large programs is very time consuming. Recently, a growing number of microcomputer software tools have become available. In many applications, these programs can be directly utilized as components of the decision support system. For example, spreadsheet programs and database programs are ideal for their appropriate simple tasks. In this section, we shall describe the implementation of the overall preference model in a decision aid by integrating different software modules. The three main parts of the system correspond to the steps (ii)-(iv) in the problem solving procedure. The overall architecture of the system is illustrated in Figure 2.

3.5.1 The Decision Analysis Module

The basic element of the integrated system is the general purpose program which performs the decision-analytic tasks: structuring, number elicitations, calculations, sensitivity analysis, etc. In our case, an interactive microcomputer based decision software package, called HIPRE (**HI**erarchical **PRE**ference analysis), was already available [5]. The package includes two possibilities for rating: the AHP technique and direct rating. The program had been developed and revised on the basis of our previous experiences in working with high level political decision makers [4]. Our research group has also developed a package for the generalization of the AHP to systems with feedback, called NETPRE (**NET**work **PRE**ference analysis) (see [5]), but the network approach was not used in the present project.

Figure 2. Overall Architecture of the Integrated System

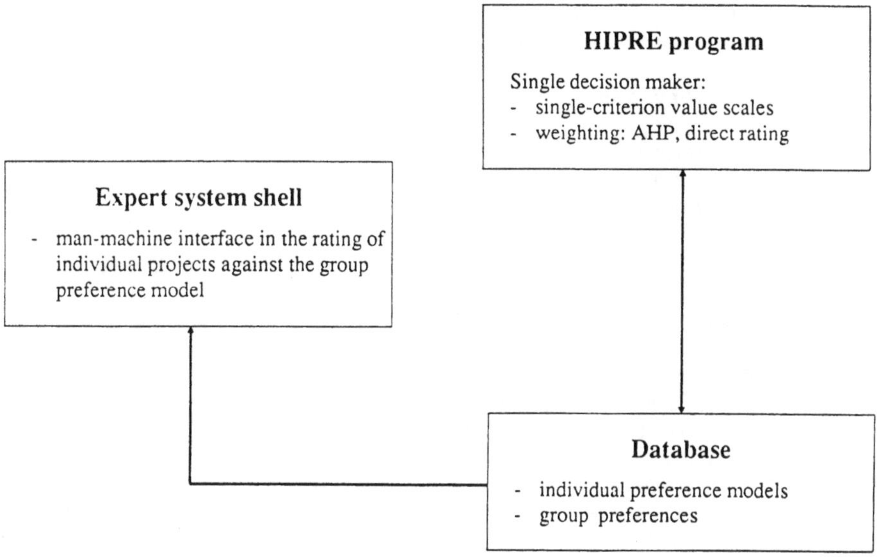

As an example of assessing a value scale with the AHP, let us consider the evaluation criterion *product's competitive environment*. It is an element of the marketing criteria under the attractiveness property of the project. Five possible values regarding the expected competition from new entrants or competitive reaction were determined:

Value	Meaning
very short	product lead will be very short
short	product lead will be relatively short
moderate	market share can be maintained
long	product lead will be relatively long
very long	a strong chance to sustain large market share.

A screen from HIPRE during an example session is illustrated in Figure 3. The complete pairwise comparisons and the resulting weights are then determined by HIPRE, e.g., as illustrated in Figure 4. It is also possible to illustrate the weights graphically (see Figure 5).

The local priorities reflect the relative position of each option with respect to each criterion. The value scale is obtained from these priorities by scaling the weights so that the minimum weight (w_{min}) is equal to zero and the maximum weight (w_{max}) is equal to one. Transformation from a relative weight w_i to a new weight s_i is given by $s_i = (w_i - w_{min})/(w_{max} - w_{min})$. From the above example we get the final weights:

very short	0.000
short	0.112
moderate	0.259
long	0.632
very long	1.000.

Let us then consider the ranking of the hierarchy under *marketing* with the HIPRE program. The three sub-criteria are *product's competitive environment*, *sales volume*, and *market trend and growth*. The single-criterion value function related to the criterion *product's competitive environment* has been constructed above. *Sales volume* is assumed to have three possible values: 60 million U.S. dollars/year, 120 million U.S. dollars/year, and 180 million U.S. dollars/year. The possible values for *market trend and growth* are rapidly declining, declining, steady, growing slowly, and rapidly expanding. The three sub-criteria are ranked by pairwise comparisons. The single-criterion value scales are then assessed as above using the HIPRE program. The results of an example session are illustrated graphically in Figure 6. The relative weights are then scaled so that the worst alternative is rated 0 and the best alternative is rated 1.

Figure 3. A Screen from HIPRE During the Assessment of a Value Scale

Level 1 Comp.env.

Level 2	1	2	3	4	5	Weight
1. very short	1	1/3	1/6	1/7	1/8	
2. short	3	1	?	?	?	
3. moderate	6	?	1	?	?	
4. long	7	?	?	1	?	
5. very long	8	?	?	?	1	

Comp.env.

a: short
b: moderate

Which is more important (a/b) ?
How much more important (1..9) ?

Comparison scale:
1 = equal importance
3 = weak importance
5 = strong importance
7 = very strong importance
9 = absolute importance

3.5.2 Link to the Database

The HIPRE program supports decision problems with a single decision maker. However, in this case, we have a group decision problem. Of course, it is possible to calculate the group preferences separately and that is what we did in this project. However, we soon found out that a more practical way of doing this is to use a decision aid that has the capability of combining group evaluations. A simple solution to the problem was to develop a database interface to the decision aid. This allows the storage of individual preference models in a database under a database program. The database program makes it easy to manage the preferences in a group setting. It also aids in comparing a new project candidate with the existing projects that have been evaluated earlier.

3.5.3 Man-machine Interface

The user interface of the HIPRE decision analysis program is very general purpose and the presence of the decision analyst is required when the aid is used. The final goal is naturally to develop a decision aid that can be used by the managers themselves. As a first step in enhancing the user-friendliness of the decision aid, a problem-specific man-machine interface was developed. The aid thus uses concepts and terms that are familiar to the company. The interface and the synthesis of the evaluations were implemented using an expert system shell. A growing number of expert system shells have become available for microcomputers. Although the capacity of present day microcomputers does not allow the implementation of large expert systems, expert system programs with rule-based declarative programming provide easy tools for implementing decision analysis models. The programmer need not program all the relationships between the system variables - only the description of the system is needed. The relationships are worked out by the inference mechanism of the expert system shell. Expert system shells also have an advanced user interface available and the user is prompted to supply input data to the system when necessary. From the available expert system shells, we chose the Personal Consultant Plus by Texas Instruments which is based on the Emycin system (see, e.g., [1]). This tool includes advanced features such as frames and metarules and it has a database interface, which is important in our application. Communication between the decision analysis program and the expert system is arranged through the database.

We shall next illustrate the use of an expert system shell in the programming of a value tree. Let us consider the programming of the interface and computations related to the element *marketing* under *project attractiveness* in the R&D hierarchy. Denote by w_i and s_i, i = 1,2,3 the local priorities and the values of the evaluation criteria *product's competitive environment*, *sales volume*, and *market trend and growth*, respectively. The value of a project with respect to marketing is then obtained with the additive model as $V(\text{marketing}) = w_1 \cdot s_1 + w_2 \cdot s_2 + w_3 \cdot s_3$.

Figure 4. The Results of the Pairwise Comparisons from HIPRE

Level 1 Comp.env.

Level 2	1	2	3	4	5	Weight
1. very short	1	1/3	1/6	1/7	1/8	0.036
2. short	3	1	1/2	1/4	1/5	0.082
3. moderate	6	2	1	1/3	1/4	0.142
4. long	7	4	3	1	1/2	0.295
5. very long	8	5	4	2	1	0.446
					C.R.	0.035

Figure 5. Graphical Illustration of the Final Weights

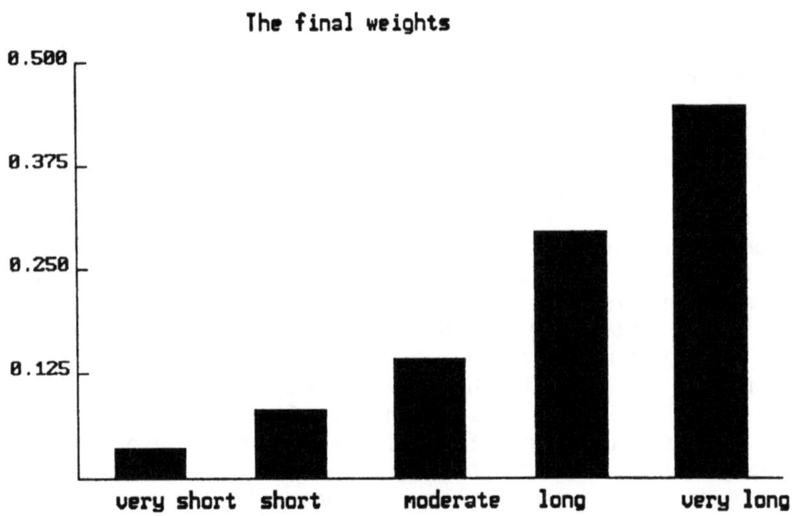

Figure 6. An Example of the Single-criterion Value Scales Assessed Using the HIPRE Program (Sub-criteria from Left to Right: Product's Competitive Environment, Sales Volume, and Market Trend and Growth)

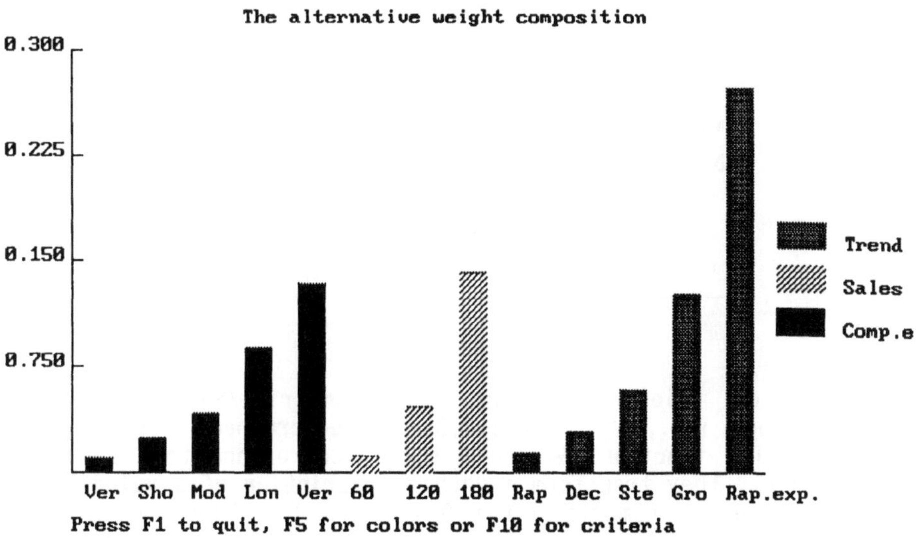

The corresponding user interface and the computations can be programmed by rules in the following way. First, define the variables that represent the lowest level criteria. The definition of a variable consists of properties. The *prompt property* determines the question that the system uses when it needs the value of the particular variable. With the *expect property* list it is possible to let the decision maker choose a value from a discrete set of given alternatives. In the rule-based program the definition of the three criteria look quite simple and readable (Figure 7).

If the value of a variable is always determined by the user, it is given the *askfirst property "Yes."* For example, when the system needs the value of *market trend and growth*, the decision maker is shown the display illustrated in Figure 8. The decision maker then selects one response from the list using the cursor control. A variable can also have a *help property*. This property can be used to clarify the question asked from the decision maker. In the above example, the decision maker can be informed about the higher level dimensions of the criterion ***market trend and growth***. With the attached help property, this information is obtained by pressing a function key (see Figure 9).

In the program, the variable ***marketing*** is defined in terms of the previously described lowest level criteria. In the knowledge base, the definition consists of a method that defines the way of computing the value of the variable:

Marketing
Method:
$w_1 * V_1$(Products–comp.–environment)+
$w_2 * V_2$(Sales–volume) +
$w_3 * V_3$(Market–trend–and–growth)

The local priorities of the evaluation criteria with respect to ***marketing***, i.e., w_i, i = 1,2,3, are determined in the HIPRE program. The functions V_i, i = 1,2,3, correspond to the value scales, i.e., they transform a verbal description of a value of a criterion into a numeric value $0 \leq s_i \leq 1$. When the system determines the project's value with respect to marketing, it begins to search values for the variables representing the three operational criteria. Since each of the three variables has the askfirst property *"Yes"*, the decision maker is asked to evaluate the project with respect to the three criteria. The value for the marketing criterion is then computed by applying the ***method property***, which in this case means using an additive preference model.

4. WORKING WITH THE MANAGEMENT GROUP

After the hierarchy had been completed and accepted by the contact person from the company, ranking the elements in the hierarchy was carried out with 18 managers representing different operations in the company (production, marketing, research and development, financing, and strategic planning). Moreover, since only one person from the company had been involved in the development of the hierarchy so far, we also wanted to obtain comments and feedback from the other managers. Five real project proposals were used to test the hierarchy. These projects represented different central business areas of the company.

Figure 7. Definition of Three Criteria in the Rule Base

Product's–competitive–environment
Prompt: What is the expected
product lead ?
Expect: Very short
Short
Moderate
Long
Very long
Askfirst: Yes

Sales–volume
Prompt: What might the expected sales
of the product be ?
Expect: Positive number
Askfirst: Yes

Market–trend–and–growth
Prompt: What are projections of the
market demand for the product ?
Expect: Rapidly declining
Declining
Steady
Growing slowly
Rapidly expanding
Askfirst: Yes

Figure 8. Display to Evaluate the Future Markets for the Product

```
                         Research and development

 ---------------------------------------------------------------------
‡What are projections of the market demand for the product ?
‡
‡   RAPIDLY_DECLINING
‡   DECLINING
‡   STEADY
‡   GROWING_SLOWLY
‡   RAPIDLY_EXPANDING
‡
‡
‡
‡
‡
‡
‡
‡
‡
‡
‡
‡  1. Use the arrow keys or first letter of item to position the cursor
‡  2. Press RETURN/ENTER to continue
‡
 ---------------------------------------------------------------------
```

Figure 9. Use of the Help Property

```
                         Research and development

 ---------------------------------------------------------------------
‡What are projections of the market demand for the product ?
‡
‡   RAPIDLY_DECLINING      Help:-------------------------------------------
‡   DECLINING              ‡
‡   STEADY                 ‡Market trend and growth is one component of the
‡   GROWING_SLOWLY         ‡attribute MARKETING under the upper level
‡   RAPIDLY_EXPANDING      ‡attribute PROJECT ATTRACTIVENESS.
‡                          ‡ ** End - RETURN/ENTER to continue
‡                          ‡
‡                          ‡
‡                          ‡
‡                          ‡
‡                          ‡
‡                          ‡
‡                          ‡
‡                          ‡
‡                           -------------------------------------------
‡
‡  1. Use the arrow keys or first letter of item to position the cursor
‡  2. Press RETURN/ENTER to continue
‡
 ---------------------------------------------------------------------
```

Following the instructions of the analyst, the managers started the ranking process by working from the general top level criteria to the lower ones. The meaning of each criterion was clarified during the analysis session and it was typical that some of the priorities were re-evaluated. This experience suggested that it may be better to start the ranking process from the bottom level in order to clarify the ranges of the criteria. The initial hierarchy was found to correspond well with the decision makers' view of the problem. However, there were criteria that did not make any noticeable distinction between the alternatives and, thus, could be ignored. For example, *availability of money* turned out to be such a criterion; all projects scored about the same. As far as the ranking method is concerned, the combination of the AHP and direct rating was used. The choice of the method was given to the managers.

With real project proposals, most members of the management group considered this phase of the project very useful and interesting. During the ranking process, the decision maker must be clear on the exact meaning of each criterion in the hierarchy. This gave rise to a lively debate among the various managers. Among other things, they wanted to ascertain the overall strategy of the company. Moreover, managers from different divisions had to evaluate the projects from the point of view of other managers. This helps in the consensus building process. The results of the ranking were delivered to the 18 participants in the form of averages and standard deviations of local and global priorities. The results were also presented to the chief executives of the company.

Since the comparisons that generate the weights at the various levels of the hierarchy are done independently by the members of the management group, it is possible that different evaluators do not develop the same kind of rankings. This is called a conflict. Since the preference models reside in a database, it is possible to point out the sources of conflict and thereby facilitate a compromise ranking. Preference models are a clear formal approach to enhance communication between the decision makers and they focus attention on the precise sources of conflict. The average values of the weights were used to represent the group's value system because of the very collaborative nature of the decision making problem with complete sharing of information. In the context of the AHP, Saaty has shown that the weights of individual decision makers should be combined by taking the geometric mean of the weights. This is due to the ratio scale of the AHP. Because of the interval scale used in the single-criterion evaluations, the arithmetic mean was used in our project.

The resulting compromise preference model clearly shows the most important evaluation factors of R&D project candidates in the company. The managers evaluate the candidates against the predetermined compromise preference model, which is implemented by the expert system shell. Differences in the rating of a candidate are thus due to differences in the evaluation of the candidate's performance with respect to the individual lowest level criteria.

The weights in the company's preference model are updated when necessary (for example, changes in the environment might require a new assessment of the relative importance of the criteria).

5. CONCLUSIONS

The paper describes a real application of an integrated decision aid for long-term strategic R&D project selection. The aid was developed for top level managers in a large Finnish company with considerable R&D investments. The results of this project support our earlier positive experiences of the potential usefulness of decision aids in strategic decision making. The main advantages were in the structuring of the problem and the resulting improvement in group communication. The decision analytic approach corresponded well to management's way of thinking.

A hierarchical representation was used for the individual preference models. The weights in the hierarchy as well as the subjective value scales of the lowest level criteria are updated by a general purpose decision analysis program which uses the AHP or direct rating as the ranking method. The preference models are stored in a database which is managed by a database program. The problem specific man-machine interface of the aid is implemented by an expert system shell. This kind of an integrated approach turned out to be very efficient from the point of view of programming. Each of the modules is ideal for its task. A technical conclusion is that the expert system approach provides many new possibilities in the implementation of problem-specific decision aids.

From the technical point of view, an important factor still affecting the wider use of decision aids among top level managers is the quality of the user-interface. The available computer software packages running the decision analytic aids are too general purpose and require the presence of a decision analyst when used. Current software implementations of decision aids have been designed to promote particular decision analysis methods. However, in the future, we expect that the starting point will be the needs of the end users.

6. REFERENCES

1. A.A. Assad and B.L. Golden, "Expert Systems, Microcomputers, and Operations Research," **Computers & Operations Research, 13**, 301-321 (1986).

2. N. Baker and J. Freeland, "Recent Advances in R&D Benefit Measurement and Project Selection Methods," **Management Science, 21**, 1164-1175 (1975).

3. R.H. Becker, "Project Selection Checklists for Research, Product Development and Process Development," **Research Management, 23 (5)**, 34-36 (1980).

4. R.P. Hamalainen, "Computer Assisted Energy Policy Analysis in the Parliament of Finland," **Interfaces, 18 (4)**, 12-23 (1988).

5. R.P. Hamalainen and R. Karjalainen, "NETPRE - A Decision Support System for Analyzing Preferences in a Network Setting," **Proceedings of the International Symposium on the Analytic Hierarchy Process, Tianjin, China**, 399-405 (1988).

6. R.P. Hamalainen and T. Seppalainen, "The Analytic Network Process in Energy Policy Planning," **Socio-Economic Planning Sciences, 20 (6)**, 399-405 (1986).

7. R.L. Keeney and H. Raiffa, **Decisions with Multiple Objectives: Preferences and Value Tradeoffs**, John Wiley & Sons, New York, New York (1976).

8. M.J. Liberatore, "An Extension of the Analytic Hierarchy Process for Industrial R&D Project Selection and Resource Allocation," **IEEE Transactions on Engineering Management, EM-34**, 12-18 (1987).

9. Neste Corporation, **Annual Report**, (1987).

10. T.L. Saaty, "Absolute and Relative Measurement with the AHP: The Most Livable Cities in the United States," **Socio-Economic Planning Sciences, 20 (6)**, 327-331 (1986).

11. T.L. Saaty, **The Analytic Hierarchy Process**, McGraw-Hill, New York, New York (1980).

12. D. von Winterfeldt and W. Edwards, **Decision Analysis and Behavioral Research**, Cambridge University Press, Cambridge, England (1986).

WATER RESEARCH PLANNING IN SOUTH AFRICA

L. Paul Fatti
Department of Statistics
University of the Witwatersrand
P.O. Wits
Johannesburg
2050 South Africa

ABSTRACT

This paper describes a method for identifying research needs and obtaining priorities for them. The overall method consists of two parts. The first is concerned purely with the identification of research needs without regard to their relative importances or urgencies. The second part is devoted to obtaining priorities for the identified needs via the Analytic Hierarchy Process. A case study is presented in which this method was used to develop a Master Plan for Research in surface hydrology and surface water resources for South Africa.

1. INTRODUCTION

South Africa is a country rich in its supply of manpower, land and minerals, but whose development is limited by its scarce supply of water. Even after the current political troubles are resolved, the water shortage problem will remain and the country's ability to solve this problem will be a prime determinant of its economic development. This ability will depend on how much is known about all aspects of water in South Africa: from its occurrence, circulation and spatial distribution, to the country's need for water and how to satisfy this need most efficiently.

In order to learn more about the problem, the Water Research Commission was established in 1971 with two primary functions: (1) coordinate, promote, and encourage the water research effort in South Africa and (2) fund a portion of the water research effort (see [6]). The Commission's structure for coordinating water research was and still is organized around a number of broad research fields as indicated in Figure 1. Each field is managed by a senior advisor, whose task it is to plan, promote, and coordinate research in that field. Each broad research field is further divided into a number of research areas. Figure 1 illustrates the case for the research field Surface Hydrology and Surface Water Resources.

Each of these research areas is then divided into a number of components and, on the next level, the individual facets of research are related to the research components. This is shown in Figure 2 for the research area Sediment and Solute Yield.

In order to help achieve its first function and provide guidance in its second, the Water Research Commission produces master research plans in each field. These plans contain a list

Figure 1. The Water Research Commission's Structure For Coordinating Water Research

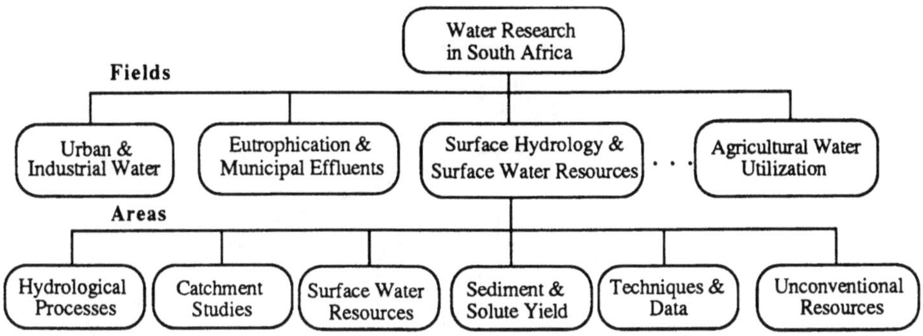

of research objectives and their priorities. They focus on the type of research that should be undertaken in each field and establish its urgency or importance. They are used to guide organizations (not only the Commission) in their decisions on the planning and funding of research.

These master research plans need to be updated from time to time in order to keep up with current research and with the changing circumstances and needs of society. During an 18-month period from 1985 to 1986, the author was involved as consultant to a small working group charged with the task of developing a new master plan for research in Surface Hydrology and Surface Water Resources. Essentially, this area focuses on the properties, distribution, and circulation of water on the surface of the land, in the soil and underlying rocks, and in the atmosphere. After some discussion on various alternative approaches towards carrying out its task, the working group settled on the approach that is presented in the remainder of this paper.

Figure 2. Components and Facets of Research Area 4

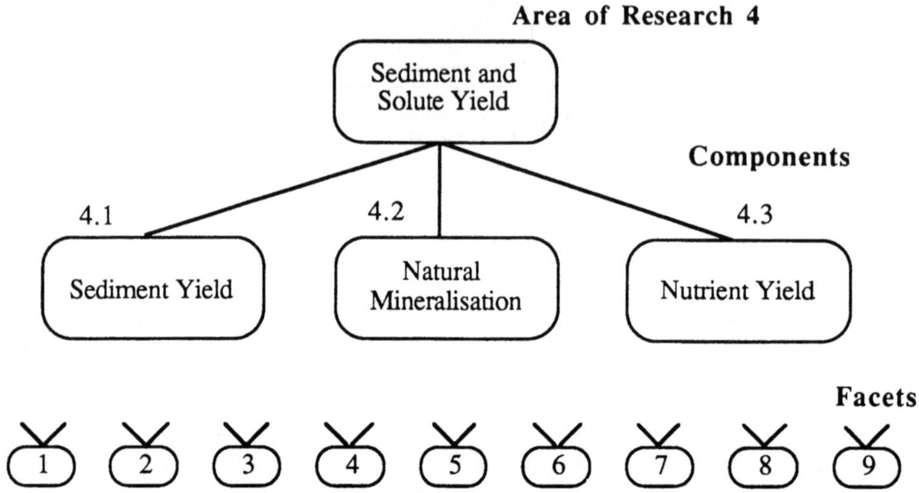

1. Detachment Transport and Deposition
2. Identification of Source Areas
3. Regional Studies
4. Land-Use and Management Effects
5. Remedial Measures
6. Sediment/Nutrient Relationships
7. Impact on Receiving Systems
8. Interactions within the System
9. Natural Rate of Solution from Parent Material

2. THE BASIC APPROACH

Two important aspects of research planning concern the identification of research needs and the assessment of their priorities. The approach proposed in this paper reflects these dual tasks. The first is concerned purely with the identification of research needs without regard to their relative importances or urgencies, while the second is devoted to obtaining priorities for the identified needs via the AHP.

While much of the approach was developed during the author's involvement with the Water Research Commission, the method has its basis in earlier work on research planning in regional development, reported by Fatti [3,4].

2.1 Identification of Research Needs

The approach is based on the interactions between the various goals (see Table 1) of the research funding agency and the different areas of research, identified by means of two-way tables, in which the goals are listed along one axis, and the research areas and their components are listed along the other axis. An individual cell in the table thus represents the interaction between a particular goal and a particular research area (or component thereof). The entry serves to stimulate ideas on research which needs to be done in that research area in order to help achieve the particular goal of the agency.

The purpose of this phase is to stimulate ideas, and no account is taken of the relative priorities or urgencies of these identified research needs. An important point to note is that some of the cells in the two-way tables may suggest several different needs for research, while others may yield none. Furthermore, the same research need may emanate from different cells in the tables.

Hopefully, at the end of this phase, we have an exhaustive list of research needs that are categorized by research area and by the research goal(s) to which they relate. In most circumstances it makes sense to consolidate these identified needs into a smaller number of distinct research objectives, or projects, each of which is expected to contribute to the fulfillment of one or more of the research goals. The next phase requires that these identified objectives be ranked in importance.

2.2 Assessing Priorities for the Identified Research Objectives

The priorities of the research objectives depend on the relative importances of the goals to which they relate. Thus, as a first step, priorities must be obtained for these research goals. The AHP is used to obtain these priorities.

Table 1. The Goals of the Water Research Commission

OVERALL GOAL	AID DEVELOPMENT AND MANAGEMENT OF THE SURFACE WATER RESOURCES OF SOUTHERN AFRICA		
PRIMARY GOAL	1.	Assess the temporal and spatial characteristics of surface water resources.	ASSESS
SECONDARY GOALS		1.1 Better understand the hydrological cycle for purposes of prediction.	UNDERSTAND
		1.2 Obtain enough data to make assessment at the desired level of accuracy.	DATA
		1.3 Develop techniques for extrapolating hydrological information in both time and space.	EXTRAPOLATE
		1.4 Develop efficient hydrological information systems.	INFORMATION SYSTEMS
PRIMARY GOAL	2.	Determine patterns and trends for the prediction of future water supply and demand as well as the relationship between these and the development of other resources.	PREDICT
SECONDARY GOALS		2.1 Develop means of predicting man's influence on the hydrological regime.	MAN'S INFLUENCE
		2.2 Better forecast the components of the hydrological cycle in the medium and long term.	FORECAST
		2.3 Develop an adequate knowledge of the inter-relationships between demographic, socio-economic, and water supply and demand factors.	INTER-RELATIONSHIPS
		2.4 Determine the influence of changing technology on future water supply and demand.	CHANGING TECHNOLOGY
PRIMARY GOAL	3.	Evaluate alternative water resource management options.	EVALUATE ALTERNATIVES
SECONDARY GOALS		3.1 Study the hydrological consequences of rural and urban land-use for the assessment of catchment management options.	CATCHMENT MANAGEMENT
		3.2 Develop more effective operating strategies for water supply systems.	OPERATING STRATEGIES
		3.3 More effectively control water losses from hydrological and/or water resources systems.	CONTROL LOSSES
		3.4 Develop and evaluate the technology to increase water supplies.	INCREASE SUPPLIES

Figure 3. The Hierarchy of Primary and Secondary Goals Used by the Water Research Commission

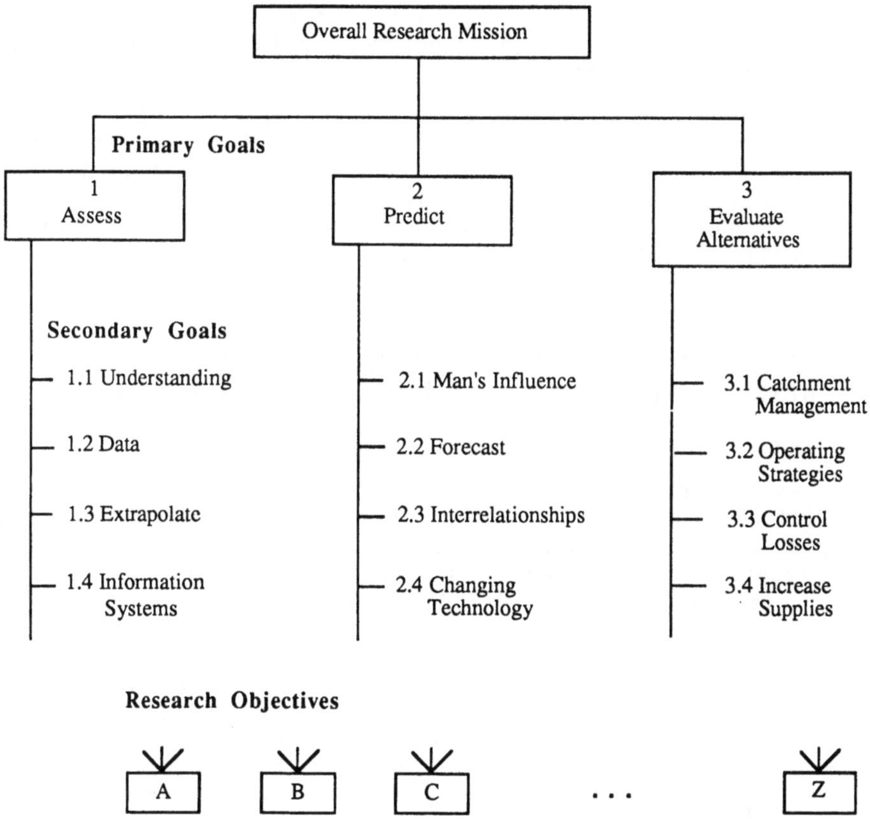

Figure 3 displays the hierarchy of primary and secondary goals used by the Water Research Commission, together with the research objectives at the bottom of the hierarchy, each relating to one or more of the secondary goals. The idea is that the AHP is used to obtain priorities for the primary and secondary goals and that priorities for the research objectives are then assessed separately with respect to each of these secondary goals. The final priorities are obtained in the usual way by weighting the priorities of the research objectives by those of the secondary goals to which they relate.

A problem in using the AHP is that the number of identified research objectives is generally too large to be efficiently compared to each of the secondary goals. In the case of the Water Research Commission, this problem was overcome by rating the research objectives according to a five-point scale, separately for each of the secondary goals, after which the overall priorities were calculated in the usual manner.

3. DEVELOPING A MASTER PLAN FOR RESEARCH ON SURFACE HYDROLOGY AND SURFACE WATER RESOURCES: A CASE STUDY

3.1 Identifying the Needs for Water Research

As described in Section 2.1, the research needs were identified from two-way tables, in which the research goals are displayed along one axis and the research areas, their components, and facets along the other axis. A separate two-way table was then drawn up for each of the six research areas. The cells were intended to stimulate ideas for research from the interactions between the individual research components or facets and the individual secondary research goals. The two-way table for identifying research needs in the area of Unconventional Resources is given in Figure 4. Five examples of research needs have been identified in this figure. In Figure 5, these research needs are described on the standard form used by the respondents in the study.

The six research areas were then divided among six members of the working group and each one was asked to identify a list of hydrological experts who might identify research needs in his particular area of research. After these lists were discussed and ratified by the entire working group, the two-way tables were sent out as questionnaires to each person on the list. Examples of identified research needs were included in the questionnaires, together with guidelines on how to record and describe each identified research need.

When the questionnaires were returned (there were approximately fifty people on each of the six lists, with a certain amount of overlap between them, and the response rate varied between 38% and 52%), the coordinators for each research area collated the responses. Thereafter, the entire working group

Figure 4. Identifying Research Needs for Unconventional Resources (Area No. 6)

COMPONENTS	FACETS	GOALS											
		1				2				3			
		1.1	1.2	1.3	1.4	2.1	2.2	2.3	2.4	3.1	3.2	3.3	3.4
6.1 WEATHER MODIFICATION													
6.2 EXTRACTION OF MOISTURE FROM THE ATMOSPHERE	HYGROSCOPIC MATERIALS								1				
	CONDENSATION												
	FOG INTERCEPTION (SCREENS)												
	FOG INTERCEPTION (VEGETATION)												2
6.3 ICEBERG EXPLOITATION													3
6.4 RAINFALL HARVESTING	IMPERVIOUS SURFACES												
	DECREASED INFILTRATION												4
	MICROCATCHMENT MANAGEMENT												
6.5 EVAPORATION AND TRANSPIRATION SUPPRESSION	ANTI-TRANSPIRANTS												
	GENETIC MANIPULATION												
	MICRO-CLIMATE MODIFICATION												
	REFLECTANT LAYERS											5	
	MONOMOLECULAR LAYERS												
	AQUATIC PLANTS												
	HYDROPHILIC COMPOUNDS												
6.6 RECLAMATION													
6.7 DESALINATION													

Figure 5. Standard Form for Respondents

RESEARCH NEEDS

* Complete the attached matrix and identify research needs by placing a number in the appropriate cell.
* Briefly describe each identified research need.
* A number may give rise to more than one need and one need may relate to more than one interaction (cell).
* Examples of research needs relating to the numbers already indicated in the matrix are given below. Please complete this form as per the examples by adding all the other research needs which you can identify.

Unconventional Resources (Area No. 6)

COMPONENTS	OBJECTIVES	IDENTIFIED RESEARCH NEEDS
6.2	2.4	**Hygroscopic materials** Evaluation of latest solar energy technology to reduce the costs of extraction of atmospheric moisture using hygroscopic materials.
6.2	3.4	**Fog interception (vegetation)** To evaluate the possibility of using trees as condensation screens in mist belts which would enable trees to maintain themselves.
6.3	3.4	**Iceberg exploitation** To evaluate the state-of-the-art technology of iceberg utilization.
6.4	3.4	**Decreased infiltration** To evaluate the interaction between infiltration suppression, runoff, soil moisture and agricultural productivity using silicone compounds.
6.5	3.3	**Reflectant layers** The development and evaluation of reflectant layers to be used on open water surfaces for the reduction of incoming radiation and thus the reduction of evaporation.

edited each response, restructured them into the form of research objectives, coordinated across research areas to avoid duplication, and identified each of the secondary research goals to which they contributed.

3.2 Assessing Priorities for the Research Objectives

The priorities of the identified research objectives were determined directly from the priorities of the primary and secondary research goals listed in Table 1 and structured in the hierarchy depicted in Figure 3.

A panel of fifteen experts in the general field of water research was employed to assess the priorities of the primary and secondary research goals. The panel represented a reasonable spread of affiliations, ranging from researchers and academics to research managers, planners, and engineers. In order to help achieve consensus in the panel, a Delphi approach (see, e.g., [5]) was used to establish the values of the two-way comparisons among all pairs of primary and secondary goals, as required by the AHP.

For each pair of goals, the panel members recorded their judgments regarding the relative importance of one goal against another on a small, pre-printed response form. The usual nine-point ratio scale was used for these pairwise comparisons. The anonymous responses were then entered into a microcomputer and the results were projected on a screen in the form of a histogram, together with the geometric mean of the responses of all panel members. After a discussion of the results among the panelists, they decided either to accept the geometric mean as their final response or they repeated the exercise, reassessing their responses in light of the previous round. In the latter case, the geometric mean of the second round responses was used as their final rating of the relative importance of the two goals. An example of such a histogram of responses is shown in Figure 6.

In order to establish priorities for all the primary and secondary goals in Figure 3, four pairwise comparison matrices were completed by the panel, one for the three primary goals and one for each of the three sets of secondary goals. Each comparison was performed using the Delphi approach described above. The pairwise comparison matrix for secondary goals 1.1 to 1.4 is displayed in Figure 7.

A general finding from this Delphi-based exercise was that the comparisons of the panel were remarkably consistent. The small value of 1% for the consistency ratio shown in Figure 7 is typical of all the pairwise comparison matrices. Table 2 gives the priorities of the primary and secondary goals which emerged from this process.

Figure 6. Histogram of Responses

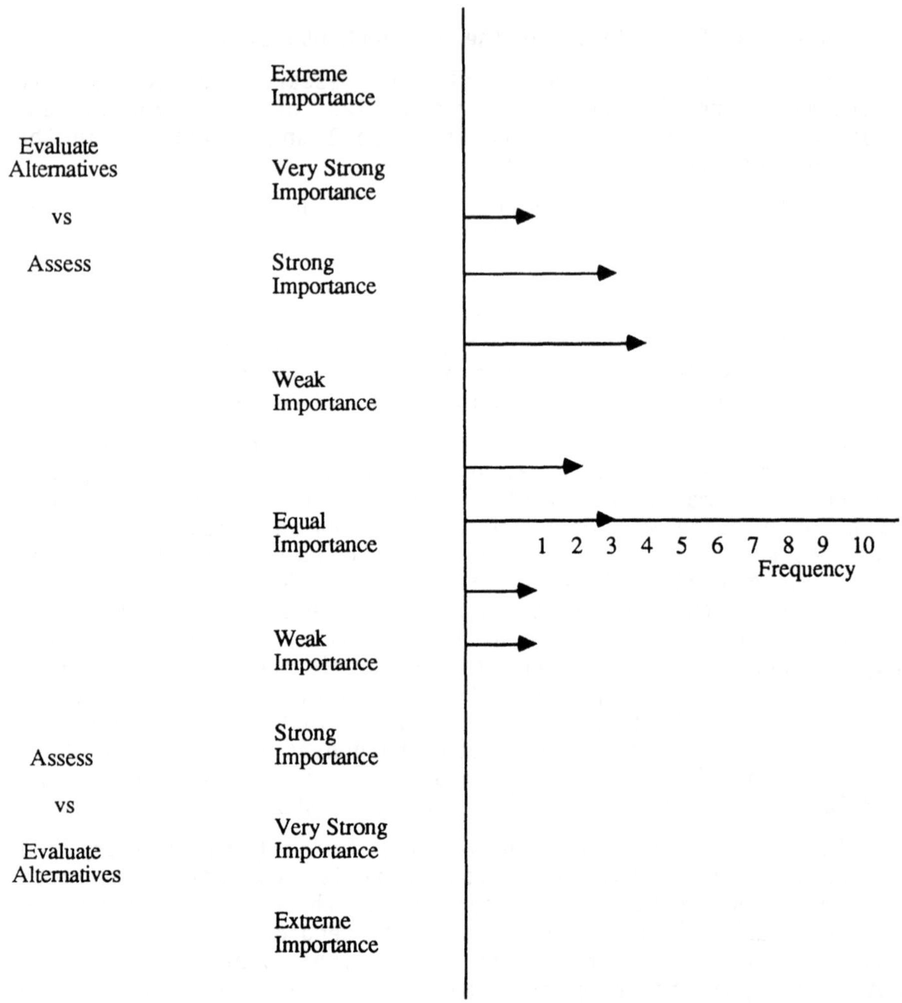

Geometric Mean
Evaluate Alternatives vs Assess
2.19

Figure 7. Pairwise Comparison Matrix for Secondary Goals 1.1 to 1.4

	UNDERSTAND	DATA	EXTRAPOLATE	INFORMATION SYST.
UNDERSTAND	1	1/1.81	1/2.06	1/2.36
DATA		1	1/1.11	1/2.04
EXTRAPOLATE			1	1/1.81
INFORMATION SYSTEMS				1

In determining the priorities of the individual objectives, we found that there were far too many (up to a hundred or more, from all six research areas, contributing to any one secondary goal) to be handled in the same way as the priorities of the research goals higher up in the hierarchy. Instead, for each secondary goal in turn, every research objective which contributed to it was rated on an A to E priority scale, depending on how important that research objective was with respect to that goal. The scale that we adopted is the same one that had been used previously by the Water Research Commission, so that it was familiar to most of the panel members. The definitions of the five priority classes on this scale are given in Figure 8.

In order to incorporate the A to E priority scale into the priority scale of the research goals, the panelists were asked to rate the A to E scale by means of the AHP. This was carried out by the panel and the results appear in Table 3. Once again, the panel's pairwise comparisons were very consistent, and it is interesting to note that the five priority classes lie along a near-perfect geometric scale with a factor of approximately 2 between the successive classes.

The actual rating of the research objectives on the A to E scale was carried out by the panelists, divided up into three groups of five, corresponding to the three primary goals. Each panelist evaluated lists of research objectives which contributed to each of the four secondary goals falling under the primary research goal to which he was assigned. There was a considerable amount of overlap between these lists, since many of the research objectives contributed to more than one of the secondary, and even primary, goals.

When the rated lists were received from the panelists, they were entered into a database, which assigned the priority scores and averaged them over the five panelists assigned to each of the three groups. Finally, priorities for all the research objectives were calculated by weighting their average priority scores by the priorities of the secondary objectives to which they related. For the purposes of the Master Plan, in which these research objectives eventually appeared, they were again categorized on an A to E priority scale.

Table 2. Priorities of the Primary and Secondary Goals as Determined by the Panel of Experts

LEVEL OF GOAL	GOAL	LOCAL PRIORITY	GLOBAL PRIORITY
PRIMARY	1	0.232	0.232
	2	0.246	0.246
	3	0.522	0.522
SECONDARY	1.1	0.135	0.031
	1.2	0.219	0.051
	1.3	0.246	0.061
	1.4	0.400	0.093
	2.1	0.283	0.070
	2.2	0.147	0.036
	2.3	0.343	0.084
	2.4	0.226	0.056
	3.1	0.233	0.122
	3.2	0.384	0.200
	3.3	0.177	0.092
	3.4	0.206	0.108

Figure 8. Priority Scale for Research Objectives

Scale	Meaning
A	An absolute priority
B	An essential priority
C	A substantial priority
D	An advantageous priority
E	A dispensable priority

Table 3. Priorities Used for Research Objectives

Scale	Priority
A	0.531
B	0.254
C	0.120
D	0.064
E	0.032

The new Surface Water Resources Master Plan (Cousens et al. [2]) has been circulated throughout the entire hydrological community in South Africa. It is being used by researchers and planners as a guide to the selection and funding of research projects. Since the Master Plan is the prime determinant of what research is carried out in this field at universities, research institutes, and by government, both the identification and ranking of the research needs has been of vital importance in determining the directions which research will take over the next decade or so. Hopefully, this will lead to a better understanding of the critical problems in surface hydrology and surface water resources in South Africa. In the final analysis, it is this enhanced understanding which will determine how South Africa solves its all-important water problems.

4. CONCLUSIONS

Our approach for identifying and ranking research needs has a number of attractive features. These include:

1. Identifying research needs in a systematic and comprehensive manner,
2. Easily matching research objectives and research goals,
3. Updating priorities in a quick and easy manner, and
4. Storing the research objectives in a database (so that it will not be difficult to produce future updates to the master research plan).

Both the identification of research needs via two-way tables and the use of the AHP for establishing priorities represent innovations in the research planning process. This successful application suggests their usefulness in the development of master plans for research in general.

5. ACKNOWLEDGMENTS

I would like to thank Mr. David Cousens, coordinator of the Water Research Commission's Working Group charged with developing the new master plan for research in Surface Hydrology and Surface Water Resources, for supporting my proposals and having the courage to implement them. I would also like to thank Mr. Cousens for making available his MBL dissertation [1]. Mrs. Kathleen Clarke of the Council for Scientific and Industrial Research deserves thanks for developing the software used in the Delphi exercise.

6. REFERENCES

1. D.W.H. Cousens, "Strategic Management: The Formulation of Objectives and Priorities with the Aid of the Analytic Hierarchy Process," Unpublished MBL dissertation, School of Business Leadership, University of South Africa, Pretoria, South Africa (1986).

2. D.W.H. Cousens, E. Braune, and F.J. Kruger, "Surface Water Resources of South Africa: Research Needs," **Water Research Commission**, Pretoria, South Africa (1988).

3. L.P. Fatti, "Approach to Identifying Research Needs in Regional Development Planning and Relating them to Research Activity," **South African Journal of Science, 79**, 184-188 (1983).

4. L.P. Fatti, "An Approach to Research Planning in Regional Development Planning," Paper presented at the 1984 International Conference on OR in Resources and Requirements in Southern Africa, 2-5 April 1984 (1984).

5. A.H. Linstone and M. Turoff, **The Delphi Method--Techniques and Applications**, Addison-Wesley, Reading, Massachusetts (1975).

6. D.F.Toerien, L.P. Vorster, and J.A. Brits, "Funding of Water Research," **South African Journal of Science, 82,** 54-56 (1986).

FORECASTING LOADS AND DESIGNING RATES FOR ELECTRIC UTILITIES

Earl R. MacCormac
Science Advisor
Office of the Governor, State of North Carolina
Raleigh, North Carolina 27603

ABSTRACT

The Analytic Hierarchy Process can be used to design rates to fulfill specific goals for ratemaking. To achieve these goals, a costing methodology must be chosen among various available types including those based upon accounting costs and marginal costs. The AHP can be employed with companies, regulators, and customers to reconcile conflicting interests in making these choices among costing methodologies. Since construction costs have a major impact upon future rates, accurate load forecasts are essential for adequate ratemaking. Again, the AHP presents a simple and rational decision aid for making such forecasts. In the cases of ratemaking and load forecasting, the AHP is much easier than traditional methods that regulators have used to make these decisions.

1. INTRODUCTION

The status of electric utilities as regulated monopolies offers an opportunity for the use of an explicit decision theory. Unlike private industries and businesses where the marketplace plays a major role in pricing, regulatory bodies set rates that are both affordable for customers and provide adequate revenues for electric companies. Regulatory commissions also usually seek to implement goals, such as conservation, stability of rates, and fairness, in their design of rates. In the past, they have additionally sought to stimulate industrial growth through rate structures.

Rates depend upon operating costs and the embedded costs of plant construction. As an extremely capital intensive industry, the construction of power plants is a major component of the costs to be recovered through rates. With the advent of nuclear power plants, the time required for construction has lengthened so that accurate load forecasts are essential. During the twentieth century, the demand for electricity has increased constantly at an average rate of 7.5% per year until the mid 1970s. This meant that the electrical capacity of the United States doubled every ten years reflecting an exponential growth curve [1]. But projections of load growth made in the late 1960s and early 1970s on the basis of assumptions that the 7.5% demand increase would continue resulted in the construction of both nuclear and fossil fuel power plants that were not needed in the 1980s when they were completed. All of the constituencies involved in electric utilities - various classes of consumers, shareholders, regulators, and managers - have acknowledged that a better method of forecasting demand is required than the simple assumption that the future will be like the past.

Many have recognized the inadequacy of present ratemaking in which regulators attempt to reach a decision that accommodates the conflicting goals, the various constraints placed upon rates, and the existence of a diversity of costing methodologies. In North Carolina, decisions about rates are made by the State Utilities Commission which receives requests for changes in rates from utility companies and then hears arguments for and against these changes in a public forum. By law, the Public Staff of the State Utilities Commission represents the interests of the public and power company points of view. The Public Staff is usually joined in its advocacy against the requests for increases by intervenors like associations of manufacturers and organized groups of environmentalists. The issues debated include what rates of return on investment the utility should be allowed, how these increased rates of return should be translated into higher rates and allocated to different classes, what cost factors should be allowed into the ratebase, and how different evaluations of the management efficiency of the company should affect the size of increases in rates.

An actual rate case can last for many weeks resulting in thousands of pages of testimony that can fill (on average) fifty to one hundred volumes of approximately two hundred pages per volume. The arguments, exhibits, and data in these volumes are very complex. Almost everyone involved in a rate case recognizes the need for simplification. Acknowledging this need, we propose to use the Analytic Hierarchy Process as a decision aid to simplify this process and resolve conflicting goals, conflicting methodologies, and conflicting actors. We have used the AHP to establish goals for ratemaking among electric utilities and regulators [2]. In 1985, we constructed a hierarchy for ratemaking and surveyed electric company executives and regulators in forty-nine states. A high degree of agreement occurred among the utilities and regulators. During this study, many utility regulators said that they would welcome the AHP as a device for ratemaking if the regulatory process could be changed to allow its use. In North Carolina and probably in most other states, the state legislature would have to pass new legislation allowing the use of a procedure like the AHP. Possible objections to such a change might arise from the feeling of the loss of due process in rate cases where everyone can be heard. Utility commissions are quasi-judicial bodies and their decisions usually can be appealed to a higher court. However, as a procedure to resolve conflicts, the AHP eliminates the need for an appeal. Adversaries who agree to participate in constructing a hierarchy tacitly acknowledge that the AHP will be the final appeal. Convincing legislatures to change the basis upon which regulatory commission decisions are made may be possible, but it will take careful argument and considerable time. In proposing the AHP as a decision aid for ratemaking, we selected a rational, effective, and simple procedure. We believe that we were successful in our use of the AHP, but we now must convince others, especially lawmakers.

In this chapter, we modify the hierarchy that was constructed in the 1985 study by Koger et al. In addition to the determination of goals for ratemaking, we attempt to decide upon a type of costing methodology that meets those goals.

The North Carolina Utilities Commission also holds public hearings before deciding upon future load forecasts. As in rate cases, the Commission hears evidence from companies whose load forecasts are disputed by the Public Staff, and possibly several intervenors. Although not as complicated as a rate case, a hearing to determine future load growth will usually encompass complicated and conflicting forecast methodologies. These include: (1) trending of historical usage patterns; (2) end-use models based on the projected stock, efficiency, and utilization of electric appliances; and (3) econometric methods that develop models to relate quantities of electricity demanded to factors such as its price, income, the price of substitute fuels, and weather [11]. After evaluating the forecasts produced by the various methods, the Commission makes a formal determination of future load growth. While the Commission is bound by confidentiality and is unwilling to acknowledge the exact forecasting method, we believe that their decision is based upon a compromise rather than the results of a single forecasting method.

We propose to use the AHP to forecast loads instead of basing decisions upon compromise among different forecasting methodologies. Our use of the AHP produces results that are similar to those currently produced by the N.C. Utilities Commission. The use of the AHP for load forecasting seems an attractive modeling possibility that deserves serious consideration by utility commissions.

Before constructing the forecasting and ratemaking hierarchies, we will sketch the constraints that form the decision-making environment in which decisions are actually made. After considering the two applications of the AHP, we will revisit their applicability to the decision-making environment implied by the constraints.

2. CONSTRAINTS UPON DECISIONS FOR ELECTRIC UTILITIES

Most rate structures are established by state regulatory commissions; bulk rates and interstate rates are established by the Federal Energy Regulatory Commission (FERC) in Washington. To focus upon the nature of constraints, we will limit our discussion to rates and forecasts determined by state regulatory bodies.

Regulatory commissions set the rate of return that a company may earn on an abstract ratebase composed of both construction costs and operational expenses. Rather than a single rate of return, there are usually several different rates of return applied to the various forms of debt - stocks and bonds, long term and short term - used to finance company costs. These rates of return are usually comparable to the rates of profit that private companies earn in the marketplace, less some discount for the fact that utilities engage in less risk since they are regulated monopolies. Rates of return are also often viewed as those rates necessary to generate stock prices that will attract investors. In setting rates of return and determining the ratebase to which these rates of return are applied, the utility regulatory commission is acting as a surrogate marketplace thereby protecting consumers from the possibility of excessively high prices that an unregulated monopoly could charge.

To determine the legitimate costs that can be included in the ratebase, regulatory commissions scrutinize an historical test year of operations, usually the year preceding that in which the case is presented. This legal constraint restricts a commission to accounting costs rather than marginal costs. As we will show later, marginal costing methodologies can be used but because of the legal requirements of looking at revenue recovery in terms of accounting costs, marginal costs can lead to the over recovery or under recovery of revenues. Marginal costs also require reconciliation in the long term or the rate structures may place one class of customers at an advantage over another.

Accounting costs consist of a complete description of all costs encountered in producing electricity. These costs include construction costs, the costs of borrowing money, and all operating costs. Marginal costs are the costs of producing the next unit of electricity without regard to previous construction or operating costs.

Regulatory commissions approve future projections of the demand for electricity and their approval is required for the construction of new generating facilities [12]. In the 1980s when many nuclear power plants, either under construction or partially completed, were abandoned, the outraged public was often ignorant of the fact that most construction plans had been approved by regulatory commissions. Through those commissions, the public had been represented in the decisions about future load forecasts and these decisions usually resulted in the approval of new power plant construction.

The limits of technology also place constraints upon decisions made concerning electric utilities. In determining the size of allowable rates of return for a particular utility, the commission takes into account the efficiency of that utility in generating and transmitting electricity and the technological limits of the equipment that affect those decisions. Until the 1970s, the marginal costs of producing electricity were falling, largely through economies of scale and improved efficiencies in turbines and generators, but by the mid 1960s thermal efficiencies stopped increasing, a fact that seemed to be ignored by both top managers of companies and utility regulators [1]. Decisions about rates of returns in the 1970s ignored the fact that the decline of marginal costs was slowing and that eventually the marginal costs of producing electricity would increase. The usual reasons given for the shift from declining to increasing marginal costs in the

1970s are the rise of fuel prices brought about by the Arab oil embargo of 1973 and the high inflationary costs of capital.

The issues of load forecasts and rates are public policy questions and take place in the political arena. By law, consumers are entitled to reliable service (uninterrupted service with little voltage or frequency fluctuation) and affordable rates that allocate costs fairly to different classes of customers. The detailed question of what constitutes an affordable rate is left to the discretion of a utility commission. Similarly, the determination of a rate structure that fairly distributes costs must be decided by the commission. As we shall demonstrate later in this paper, different costing methodologies place different classes of customers at an advantage. The question of fairness involves decisions about goals for ratemaking and then the selection of a costing methodology that fulfills these goals.

Various classes of customers bring political pressures to bear upon the utility commission by hiring lawyers to be intervenors in rate cases and by publicly presenting their cases in the media. For example, a small group of customers or opponents of nuclear power can achieve influence far beyond their numbers through media publicity. Pressure created by the media can easily influence a decision. In the 1970s, the public outrage against large increases in utility rates for residential customers was so great that many utility commissions failed to allow full recovery of costs by companies.

Decisions made by regulatory commissions should be rational and must be capable of adjudicating conflicting goals and interests without either violating or ignoring constraints. We present two examples of the use of the AHP in making regulatory decisions: (1) forecasting loads; and (2) selecting costing methodologies to achieve a series of goals for ratemaking. We have chosen four sets of actors: (1) Power companies; (2) Regulators; (3) Residential Customers; and (4) Industrial Customers. All four have different interests in both load forecasting and ratemaking. To simplify the hierarchies, we have excluded the Commercial Customer Class. This class has not been extremely vocal in recent debates about decisions concerning load and rates.

3. LOAD FORECASTING

Figure 1 represents a hierarchy for forecasting loads. Level 1 is occupied by actors concerned with projections of load growth. This growth will eventually affect rates through the absorption of construction costs into the rate base when new plants come on line. Although Power Companies and Regulators have the greatest direct concern about forecasting loads, classes like Residential Customers and Industrial Customers possess indirect concerns since they both know that growth in the demand for electricity will affect rates (until 1970, this growth meant lower rates; since then additional growth has meant increases in rates).

Figure 1. A Hierarchy for Forecasting Loads

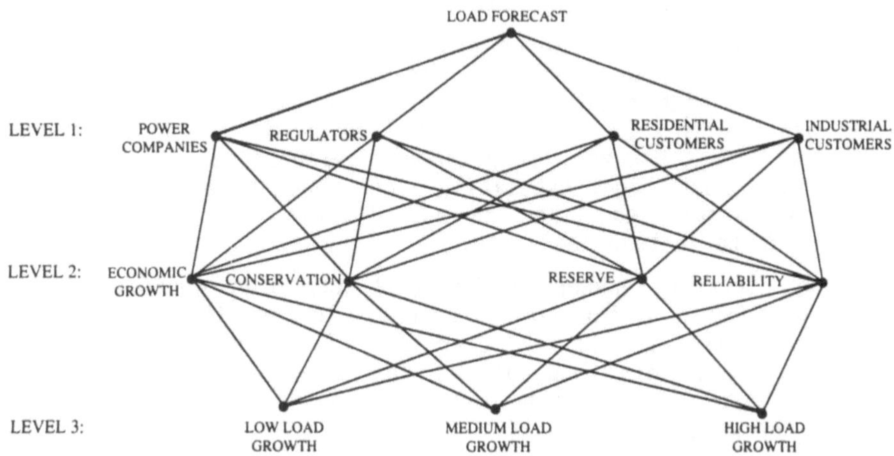

Level 2 contains four major factors affecting projections of loads. Economic Growth means additional demand for electricity but this can be partially absorbed by Conservation. Baseload generating facilities (nuclear or coal-fired plants) have become so expensive to build in the 1980s that many power companies have attempted to meet additional demands for electricity by resorting to conservation measures like load control (controls which interrupt large appliances) and time of day rates. Nuclear power plants like Oconee in South Carolina (Duke Power) were constructed in the 1960s for less than $200 per kilowatt while present costs range from $2,000 to $5,000 per kilowatt. In a period when marginal costs are rising, the construction of additional generating plants may result in lower rather than higher revenues as political pressures prevent utility commissions from passing on the full costs to the ratepayers. Instead, a commission may allow lower rates of return to the company rather than passing on the full costs of additional capacity. So many utility companies have followed this strategy that some analysts worry that in the 1990s the United States will have insufficient electrical capacity to continue economic growth [9]. With few new power plants planned for the future and with the inevitable aging and obsolescence of older power plants, the late 1990s may be a time when loads fail to increase because of the absence of capacity to meet the demand.

The reserve that a company requires refers to that excess capacity necessary to allow routine maintenance and to allow for unforeseen failures in generation. The usual reserve requirement is 20-25% but varies depending upon the mix of different sources of generation (nuclear, coal, gas, turbines, hydro) for a particular company. Reliability refers to the desire of companies and regulators to provide uninterruptible service with little voltage or frequency variation. Without sufficient reserve, reliability becomes difficult. Adequate reserve, however, does not guarantee reliability since other factors like adequate transmission facilities are also necessary. Reserve is a necessary but not sufficient condition for reliability. These two features of electric service possess enough different attributes so that they are listed separately in the hierarchy.

Level 3 shows the three possible outcomes for future load growth: (1) Low Load Growth; (2) Medium Load Growth; and (3) High Load Growth. High might be interpreted as 7.5% growth; Medium as 5.0%; and Low as 2.5%. The average load growth for 1900 through 1970 was 7.5%. However, from today's perspective, this average load growth seems high, especially since power companies have had to devise strategies like the all-electric home to stimulate consumption to sustain that rate.

The hierarchy shown in Figure 1 has been designed as a prototype for use by the various actors. To test its feasibility, the author engaged in an exercise of making the pairwise comparisons from his knowledge of the interests of the various actors. Expert Choice was used to record the comparisons and compute the weights.

Figure 2 shows the local choices of the various actors and each actor's choices of the importance of the four second level factors affecting load forecasts. Not surprisingly, Power Companies and Regulators were more concerned with load forecasts than either Residential or Industrial Customers. Both of the latter groups have come to assume that their demand will be met with reasonable rates since that has been true for most of this century. All four actors believe that Economic Growth has the greatest influence upon future load growth. Regulators express relatively high concerns for both Reserve and Reliability; by law, they must ensure adequate and uniform electrical service. Industrial Customers express more concern than Residential Customers for sufficient Reserve and Reliability because many of their industrial processes depend upon steady uninterrupted electrical service. Residential Customers express the greatest concern for Conservation as affecting future load growth since they know that the construction of additional power plants will eventually increase their rates.

Figure 3 displays the local choices of Regulators and how these choices affect the forecasts of future loads. Regulators believe that Economic Growth will result in the likelihood of a High or Medium demand for additional loads rather than Low demand. For Conservation, Reserve and Reliability, they believe that Low or Medium, rather than High Load forecasts, will result.

Figure 2. Local Choices of Various Actors

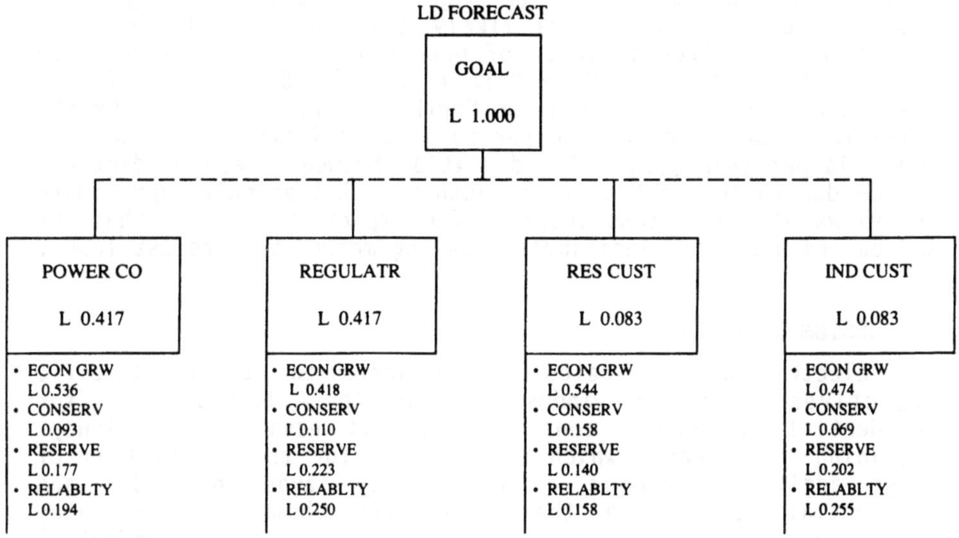

Figure 3. Local Choices of Regulators

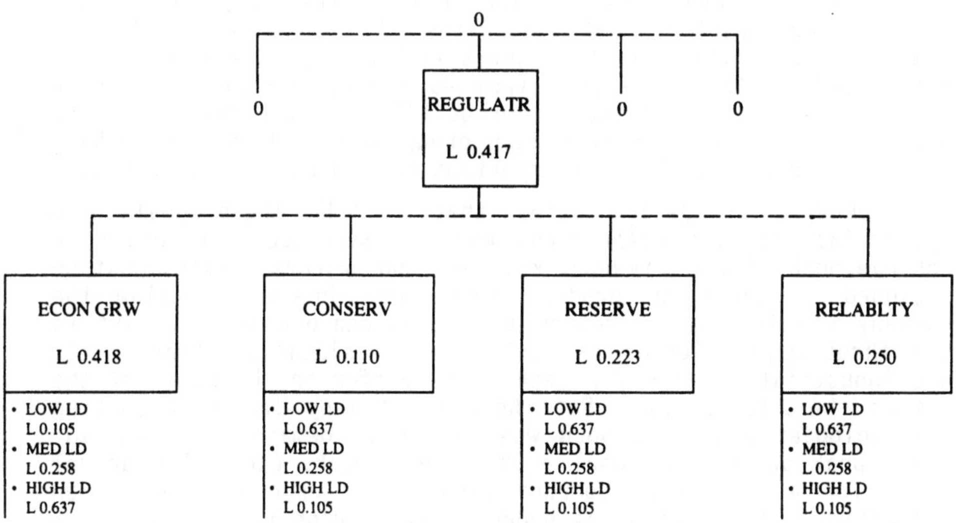

The synthesis of this hierarchy yields the following values for the three choices of future load forecasts: .380 for Low; .343 for Medium; and .277 for High. Our analysis suggests a growth rate of 2.5% as more likely than one of 5.0% or 7.5%. Combining the author's vector of preferences produces a combined growth rate of 4.74% (.380 x 2.5% + .343 x 5.0% + .277 x 7.5%). The actual forecasted growth for North Carolina determined by the Utilities Commission for the period 1986-1995 ranges between 2.2% and 2.9% per year [12]. The deviation between these predictions may be due to the fact that the author's choices only approximate the views of the various actors. The important point is that the AHP can be used as a legitimate modeling method to forecast loads.

4. DESIGNING RATES

Electric utility rates are designed to achieve specific objectives such as industrial growth, affordability for residential consumers, conservation, and economic efficiency. Declining block rates which charge less for the consumption of large additional amounts of electricity favor industries and residential users like the all-electric home. Such rates encourage industrial growth but may lead to higher residential rates for the average homeowner. Declining block rates also conflict with the goal of conservation. These conflicting goals demand a method of reconciliation that produces results which can be realized in actual rate design. We have used the AHP to reconcile the major goals as (1) Revenue Requirements; (2) Simplicity of Rates; (3) Stability of Rates; (4) Rates that further Conservation; and (5) Rates that are Fair [2]. A questionnaire based upon these goals with appropriate subgoals was circulated to 184 regulators from 49 states (we excluded North Carolina) and to 104 of the larger utilities in the United States. More than 50% responded from each group with both groups ranking Revenue Recovery as the most important goal and Fairness second.

However, to design rates that fulfill these goals, an appropriate costing methodology must be selected. In order to achieve both adequate revenue recovery and fairness, rates must be designed in order to produce sufficient income to allow the company a fair rate of return and to allocate costs "fairly" to the three major classes of customers: Residential, Commercial, and Industrial. Rates for each class differ on the basis of the demand characteristics. But there exist many different costing methodologies depending upon what types of demand are measured: daily peak demand, seasonal peak, hourly peak, etc. The actual costs for a particular company also depend upon its generation capacities. Some companies depend heavily upon nuclear power or fossil-fueled plants for their baseload while others depend upon hydroelectric power. In examining a set of generating facilities, an analyst can determine which costing methodologies favor a class of customers [3,6].

Traditional costing methodologies for electric utilities have been based upon accounting costs. Regulatory bodies filling the role of a surrogate marketplace for a regulated monopoly have attempted to determine costs based upon a past test year of operation. They determine the embedded costs of construction and equipment and the operating costs of generation, transmission, and service. They create a fictional ratebase to which they apply an agreed upon series of rates of return with respect to short term and long term debt incurred to cover these costs. Economists have argued vigorously that a more economically efficient methodology would be to employ marginal costs rather than accounting costs [8]. Vigorous debates occur over whether to use marginal or accounting costs and if the former, then what are the short term and long term results. Spot pricing based upon hourly measurements of peak demand and production costs projected to a twenty-four hour cycle has been proposed as one of the best forms of short term marginal costs [10].

The discovery that the cost of electricity cannot be determined in a precise, unequivocal and objective manner does not mean that costs, however measured, should not be a primary factor in judging the fairness of distribution. The methodological variability of the cost factor does mean that objective cost-based rates do not exist. Admitting that different "costs" can be obtained for the same production, transmission, and consumption of electricity in a single utility company raises the question of why a particular method adopted for measuring costs was selected. Such a decision can only be justified by reconciling the different objectives stated by various parties interested in rate design. For the purpose of this study, the major actors are: (1) Power Companies; (2) Regulators; (3) Residential Customers; and (4) Industrial Customers.

Four very general types of costing methodologies have been chosen for consideration: (1) Short Term Marginal Costs; (2) Long Term Marginal Costs; (3) Accounting Costs that Favor Industries; and (4) Accounting Costs that Favor Residential Customers. Short Term Marginal Costs represent various types of spot pricing. Long Term Marginal Costs represent a marginal costing methodology that employs a mechanism for reconciling revenue. Marginal costing applied to electric utilities will produce both excesses and deficiencies of revenue in the short term since capital costs are non-linear: additional electricity can be generated up to the capacity of the existing equipment, but beyond that capacity there must be a large investment to produce any additional units. The Peak and Average costing methodology is an example of Accounting Costs that Favor Industries, while Average and Excess favors Residential Customers [5].

Figure 4 represents a hierarchy designed to select a costing methodology acceptable to competing actors concerned with rates. All actors possess common objectives but with different weights placed on: (1) Revenue Recovery; (2) Simplicity; (3) Stability; (4) Conservation; and (5) Fairness. By Fairness we do not assume the existence of an objective, unequivocal costing methodology that will assign the proper shares of costs to various customer classes. Rather, we understand Fairness to mean the avoidance of undue discrimination. (We have expanded upon the AHP as an arbiter of fairness in previous work [4, 5, 7].)

Figure 4. Costing for Ratemaking: A First Hierarchy

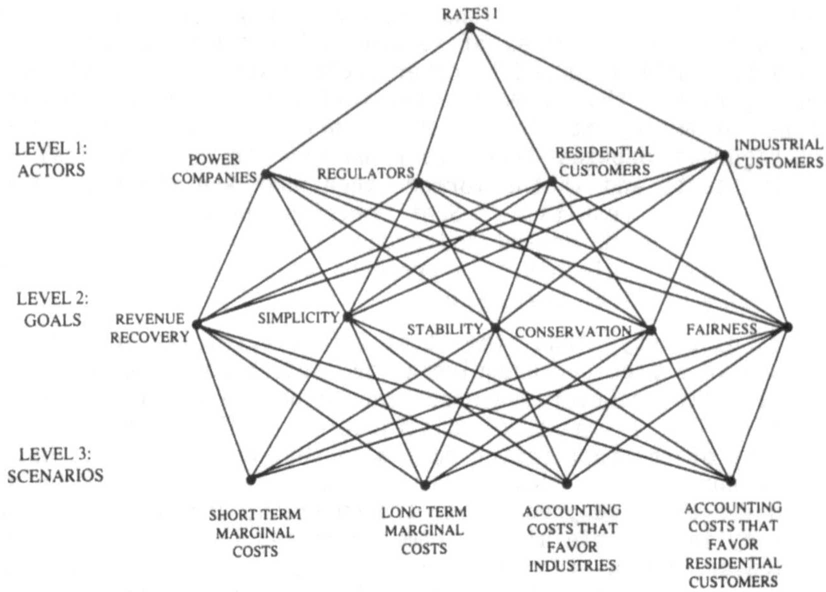

Unlike our earlier study [2], the hierarchy presented in Figure 4 has not been used with actual ratemakers. For the purpose of illustration, the author made pairwise comparisons adopting the points of view of the various actors. In the initial pairwise comparison of the actors with their concern for rates, the following vector of priorities resulted: Power Companies, .390; Regulators, .068; Residential Customers, .152; and Industrial Customers, .390. There might be some surprise at the low value of the importance of rates perceived by Regulators.

We see in Figure 5 (based upon the author's choices) that the Power Companies are concerned primarily with Revenue Recovery, .585, and secondly with Fairness, .201. Regulators express equal preferences for all five goals while both Residential and Industrial Customers express the most concern for Fairness.

Figure 5. Local Goals for Ratemaking

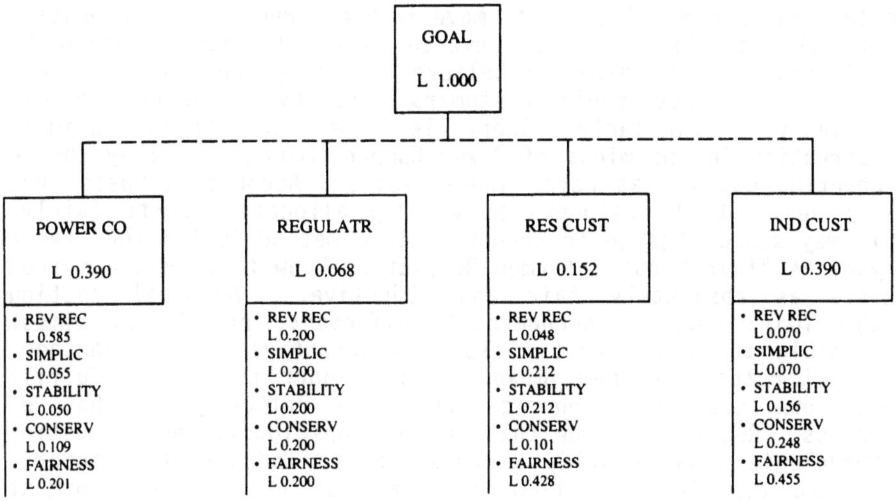

Figure 6. Local Power Company Priorities for Costing Methodologies

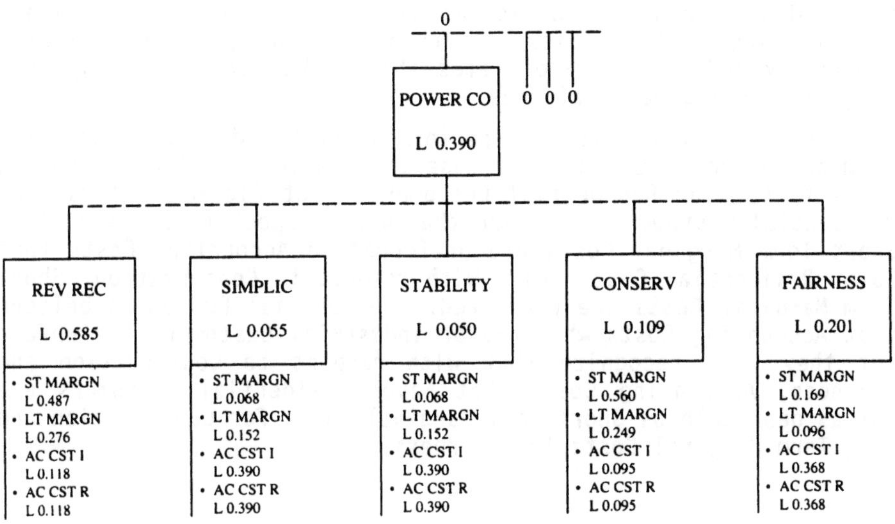

Figure 6 shows that Power Companies prefer Short Term Marginal Costs like Spot Pricing more than Long Term Marginal Costs because Spot Prices yield a greater economic efficiency while Long Term Marginal Costs move in the direction of accounting costs through devices like revenue reconciliation required by Regulators. Power Companies believe that they can maximize their profits and serve their customers best by pursuing economic efficiency. Similarly, Short Term Marginal Costs maximize Conservation in the minds of Power Company managers. They choose Accounting Costs that Favor Industries and Accounting Costs that Favor Residential Customers as equally allocating costs fairly. This may sound strange to choose costing methodologies that favor anyone as "fair," but informed Regulators know that no one method exists as absolutely fair and objective. Marginal costing methodologies may be economically efficient, but it is almost impossible to decide what class they advantage or disadvantage. Marginal rates may seem impartial, but short term rates fulfill their own goal of economic efficiency rather than that of Fairness. Regulators know that if they consciously choose costing methodologies that provide an advantage to industrial customers, they are consciously following a path of economic development hoping that in the long run, it will also provide an advantage to residential customers through additional jobs and income. If they choose costing methodologies that favor residential customers, they know that when industrial customers suffer too much, again in the long run, residential customers may lose jobs and income. Hence, both Accounting Costs that Favor Industries and Accounting Costs that Favor Residential Customers are explicitly recognized as involving conscious decisions that involve Fairness. These same costing methodologies are recognized as producing greater simplicity and stability of rates than either short or long term marginal costing methodologies.

If the same analysis is performed for Residential Customers, we might expect that in every case these actors prefer Accounting Costs that Favor Residential Customers. Yet Figure 7 shows that Residential Customers recognize that with respect to Conservation, Short Term Marginal Costs are preferred to Accounting Costs that Favor Residential Customers. With respect to Conservation, Short Term Marginal Costs are preferred. Residential Customers believe that Accounting Costs which Favor Industrial Customers are unfair but they also recognize that with respect to Conservation and Fairness, Accounting Costs that Favor Residential Customers are not as desirable as Short Term Marginal Costs for Conservation and Long Term Marginal Costs for Fairness.

Figure 7. Local Residential Customer Priorities for Costing Methodologies

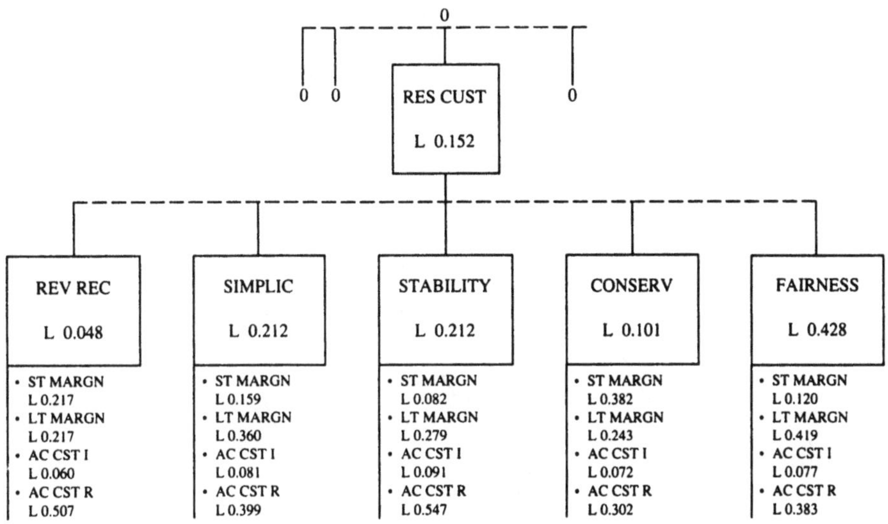

A synthesis of the hierarchy in Figure 4 yields the following vector of preferences: Accounting Costs that Favor Industries, .317; Short Term Marginal Costs, .255; Long Term Marginal Costs, .248; and Accounting Costs that Favor Residential Customers, .181. The overall Inconsistency Index for this synthesis is .04. This scenario suggests that Rates based upon Costing Methodologies that Favor Industries fulfill more of the Goals for Ratemaking of the various actors than Rates based upon either Short Term or Long Term Marginal Costs. The hierarchy suggests that Rates based upon Costing Methodologies that favor Residential Customers should not be employed. Why? Probably because such Rates may lead to economic stagnation by discouraging industrial growth thereby frustrating the various goals for ratemaking. If Rates based upon Costing Methodologies that Favor Industries disadvantage Residential Customers to the point of undue discrimination, then our notion of Fairness will be violated and the values of the hierarchy will be altered thereby changing the final choice of a costing methodology. To avoid such a consequence, the ratemaker might be tempted to choose either Short Term or Long Term Marginal Costs, especially since they are reasonably close in value to the choice of Accounting Costs that Favor Industries.

Since no single hierarchy fully captures the goals of decision makers, we now construct a similar hierarchy for ratemaking but with more general goals for ratemaking. In Figure 8, we chose the same actors and scenarios but included (1) Economic Efficiency; (2) Fairness; and (3) Regulatory

Figure 8. Costing for Ratemaking: A Second Hierarchy

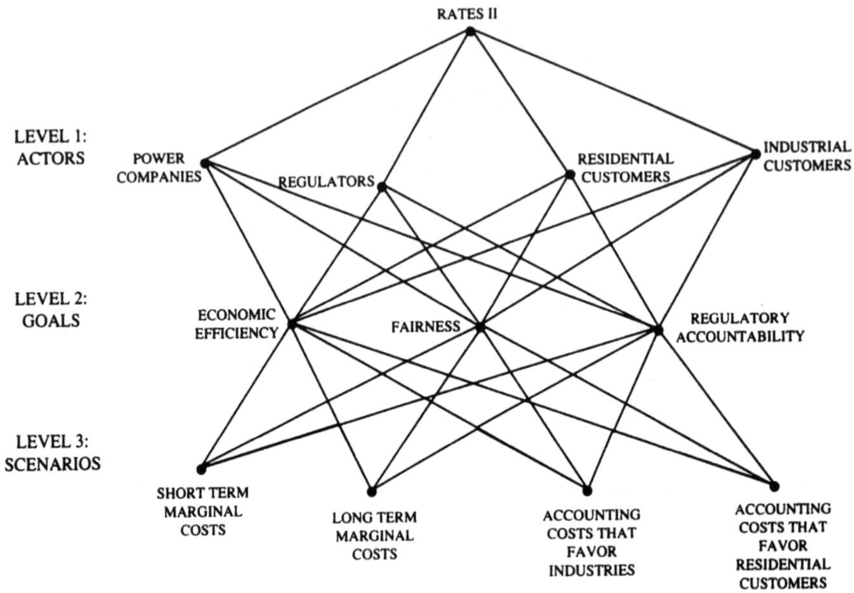

Accountability as goals for ratemaking. Economic Efficiency was chosen as a goal that benefits both the power company and the customers. An economically efficient company will generate adequate revenues and affordable rates. As in the case of the earlier hierarchy for rates, Fairness refers to the equitable allocations of costs to different customer classes. Regulatory Accountability satisfies the purpose of a governmental surrogate for the marketplace.

The synthesis of this hierarchy yields the following results: Accounting Costs that Favor Industries, .370; Short Term Marginal Costs, .253; Long Term Marginal Costs, .230; and Accounting Costs that Favor Residential Customers, .148. The overall Inconsistency Index for this hierarchy is .06. But the preferences follow the same order and are close in value. What does this tell us? The second hierarchy reinforces our belief that the AHP can serve as a legitimate instrument in selecting appropriate costing methodologies.

5. CONCLUSIONS

The use of the AHP to forecast loads and to select costing methodologies consistent with goals for ratemaking can be accomplished within the legal, technological, and political constraints which we described earlier. A state public utilities commission along with conflicting constituencies - power companies, ratepayers, and others - could decide to use the AHP to construct a hierarchy, make the comparisons, and then implement the results. Utility commissions require public participation through public hearings. No legal impediments stand in the way of holding a public discussion on a particular hierarchy that was constructed as part of the normal regulatory process of forecasting loads and designing rates.

6. REFERENCES

1. R. Hirsh, **Technology and Transformation in the American Electric Power Industry**, Cambridge University Press, Cambridge, England, forthcoming.

2. R. Koger, J. Canada, and E. MacCormac, **Decision Analysis Applied to Electric Utility Rate Design**, The National Regulatory Research Institute, Columbus, Ohio (1985).

3. E. MacCormac, "Lifeline: Equitable or Inequitable?" **Electric Ratemaking**, 1, 14-51 (1982).

4. E. MacCormac, "Values and Technology: How to Introduce Ethical and Human Values into Public Policy," in **Research in Philosophy and Technology**, P. Durbin, editor, JAI Press, Greenwich, Connecticut (1983).

5. E. MacCormac, "Ethics and Technology: Fairness in Ratemaking," in **Technology and Values: Decision Theory and Electric Rates**, J. Burnett, editor, Davidson College, Davidson, North Carolina (1984).

6. E. MacCormac, "Technological Decision Making," unpublished manuscript (1986).

7. E. MacCormac, "Werte und Technik: Wie man ethische und menshliche Werte in offenliche Planungsentscheidungen einbringt," in **Technikbewertung Philosophishce und psychologishche Perspektiven**, W. Bungard and H. Lenk, editors, Suhrkamp, Frankfurt, W. Germany (1987).

8. T. McCraw, **Prophets of Regulation**, Harvard University Press, Cambridge, Massachusetts (1984).

9. P. Navarro, **The Dimming of America**, Ballinger, Boston, Massachusetts (1985).

10. F. Schweppe, M. Caramanis, R. Tabors, and R. Bohn, **Spot Pricing of Electricity**, Academic, Boston, Massachusetts, forthcoming.

11. State of N.C. Public Staff Utilities Commission, **Analysis of Long Range Needs for Electric Generating Facilities in North Carolina**, State of North Carolina, Raleigh, North Carolina (1986).

12. State of N.C. Utilities Commission, Docket No. E-100, SUB 50, State of North Carolina, Raleigh, North Carolina (1986).

PREDICTING A NATIONAL ACID RAIN POLICY

Robert Lewis
Baltimore Gas and Electric Company
Charles Center
Baltimore, Maryland 21203

Doug E. Levy
Chesapeake Biological Laboratory
University of Maryland
Solomons, Maryland 20688

ABSTRACT

Over the last 20 years, the debate among legislators, regulatory officials, manufacturing associations, and environmental groups over an acid rain policy for the United States has yet to be resolved. The high level of disagreement among the knowledgeable participants about the complex nature of the acid rain problem and about appropriate solutions has stalled repeated attempts by the U.S. government to reach a consensus of opinion on which policy option is best. Yet, the outcome of each year's debate and the resulting policy can dramatically impact the operations of electric power companies and manufacturers of aluminum, steel, and automobiles. Faced with the current impasse, these firms still need to predict which acid rain policy might be adopted so that they can formulate an effective yearly business plan.

In this paper, we present an AHP-based model that is designed to help power company decision makers predict a national acid rain policy. We envision our model and the accompanying analysis as forming a decision support template that can be regularly updated by management to gain new insights about acid rain policies under consideration. The results of this modeling exercise can be used by power companies to help plan capital budgeting decisions related to the timing and design of new electric power plants, as well as to specify required "clean air" modifications to plants already in operation.

1. INTRODUCTION

The debate over an acid rain policy for the United States is focused on the issue of whether additional federal legislation is required to further reduce current and future emissions of sulfur dioxide (SO_2) and nitrogen oxide (NO_x) into the atmosphere. These emissions originate in U.S. factories, motor vehicles, and electric power plants when coal, oil, and fossil fuels are burned and they fall back to earth as dissolved materials in rain and snow. In the past, knowledgeable experts and special interest groups either vigorously supported or opposed increased national emission controls for a variety of reasons. As a result, conflicting conclusions based on scientific and economic evidence, as well as on political factors, made it difficult for the U.S. government to determine an appropriate acid rain policy.

Selecting an acid rain policy is an important decision problem with far-reaching implications in the public and private sectors. Of major concern is the environmental damage to lakes, streams, and forests located in the eastern United States and Canada that has been linked to acid rain [15]. In particular, the uncertainty surrounding the long-term consequences of this damage motivates the environmentalist's cautious position regarding acid rain. Environmental groups believe that the U.S. should quickly adopt a policy that dramatically reduces emission levels from existing factories and power plants.

Other groups, such as power companies and manufacturers, believe that current legislation sufficiently controls emissions. They cite an EPA study that reports a reduction of 21% in SO2 levels and 11% in NOx levels between 1976 and 1985 [8]. These groups also argue that new clean coal technologies will further reduce emissions during the 1990s. Of great concern is the cost associated with adopting a strict acid rain policy. The Congressional Budget Office estimates that reducing SO2 emissions by 50% of the 1980 level will add $1.9 to $2.1 billion to annual electricity costs by 1995 [19]. Adopting a strict policy might also lead to tax hikes, unemployment increases, and high rates of inflation. In particular, manufacturers fear that a strict policy will increase equipment and electricity costs (they must now purchase pollution control equipment for their factories) which in turn might lead to higher prices for their products. In the long run, U.S. products might be less competitive in world markets.

Our analysis will focus on three policy alternatives that the U.S. can adopt with respect to acid rain. The **Strict Nationally Mandated Controls** alternative would require substantial emissions reductions in coal burning facilities over the next 5 to 7 years. The reductions would be achieved by installing flue-gas desulfurization units (commonly called wet-scrubbers) at power plants and factories. Wet-scrubbers remove 80% to 90% of the sulfur dioxides from combustion by-products by spraying lime or limestone through the emissions prior to release into the atmosphere. Although the actual wet-scrubbing process is quite simple, the units are expensive to install and they can decrease the operational efficiency of a power plant or factory.

The second alternative, known as **Increased but Flexible Controls**, allows manufacturers and power plants flexibility in meeting stricter emission standards by using clean coal technology. Currently, about a dozen technologies are available that are designed to help coal burn more efficiently thereby producing less harmful emissions. For example, fluidized bed combustion burns coal at a very high temperature that fully oxidizes any SO2, while coal gasification converts coal into a gas that burns efficiently and does not produce SO2.

The third alternative, **Maintain the Status Quo**, would rely on the current Clean Air Act (originally ratified in 1970 but amended over the last nineteen years) and state regulations to control emissions.

As mentioned earlier, the ongoing debate about selecting an acid rain policy includes input from a variety of participants (e.g., legislators, regulatory officials, manufacturing associations, and environmental groups) who annually propose a wide range of options. Given the complex issues associated with this problem and a high level of disagreement among the knowledgeable participants, the U.S. government has yet to reach a consensus opinion on which option is best. However, the outcome of each year's debate and the resulting policy can dramatically impact a firm's operations (e.g., national legislation could require installation of wet-scrubbers that cost millions of dollars). Faced with the current impasse, many firms still need to predict which policy might be adopted so that they can formulate an effective yearly business plan. For example, consider an electric utility company that generates power using a mix of fuels--coal, oil, and natural gas. The utility uses a combination of high- and low-sulfur coal and some of its plants are fitted with wet-scrubbers. New legislation that requires the company to use only low-sulfur coal or outfit all plants with wet-scrubbers would have a substantial economic impact on the company and its customers. Clearly, this legislation affects the selection and timing of capital investments such as adding new electric power plants. It also has an impact on the design of new stations as well as on distribution decisions related to coal supply.

The present analysis proposes to model the acid rain policy problem using the Analytic Hierarchy Process. The hierarchy is constructed from the viewpoint of an electric utility that must formulate a business plan in light of the uncertainty surrounding the acid rain problem. Its goal is to identify which of the three alternatives (i.e., Strict Controls, Flexible Controls, and Status Quo) will be adopted by the U.S. government in the forthcoming year. We envision our hierarchical model and the accompanying analysis as forming a decision support template that can be updated by management on a regular basis to gain revised insights and generate new predictions about acid rain policies under consideration.

In the next section, we provide a brief history of U.S. legislation that was designed to lower emissions. The third section develops the decision model and presents results using data gathered during the Reagan Administration. The paper concludes with a discussion on how this model might actually be used by an electric utility company.

2. BACKGROUND OF LEGISLATION

The first major national legislation in the United States that resulted in actions which reduced SO2 and NOx emissions was the Clean Air Act of 1970. A time line of events related to this act (e.g., key amendments) is shown in Table 1. This act contains

Table 1. Acid Rain Time Line

Date	Event
1970	Clean Air Act is ratified--national framework for air quality standards is set. New Source Performance Standards (NSPS) take effect.
1974	Energy Crisis amendments to the Clean Air Act. Amendments stimulate coal use, relax some standards, and extend original 1975 deadline to 1979.
1977	Clean Air Act is amended. Auto emission control deadlines extended for economic reasons and NSPS revised and toughened.
1980	Congress passes the Acid Precipitation Act--forms Interagency Task Force on Acid Precipitation which begins conducting National Acid Precipitation Assessment Program (NAPAP).
1981	Formal negotiations to develop a joint U.S.-Canada agreement on transboundary pollution take place. Governments agree to promote vigorous enforcement of existing laws. Clean Air Act authorizations expire. EPA continues funding and continuing resolutions extend act's funding until 1982.
1985	Reagan-Mulroney Summit takes place. Special Envoy Drew Lewis assigned to work with William Davis to make report to President.
1986	Lewis/Davis Report submitted. Reagan Administration acknowledges acid rain is a problem for the first time. Reagan commits $5.5 billion to continued research and demonstration of new pollution control techniques.
1990	NAPAP final report due.

seven provisions that are designed to improve the air quality in the U.S. These provisions are shown in Table 2. Over the last two decades or so, the requirements of the Clean Air Act have forced many firms to convert coal burning power plants to "cleaner-burning" oil plants. Other firms substituted low sulfur coal in some production processes (this "compliance" coal produces low SO2 emissions when burned). Still others added wet-scrubbers to existing plants. However, power plants that were constructed prior to 1971 were exempted from the regulations of the act.

In 1977, Congress revised and toughened the New Source Performance Standards (NSPS) provision of the Clean Air Act. The new standards are listed in Table 3 and they apply to all plants constructed after 1978. From this table, we see that the major difference between the 1971 and 1978 standards lies in the use of low sulfur coal--the earlier standards permitted this coal to be used as a pollution control while the revised provisions did not [19]. The 1978 standards effectively mandated the installation of wet-scrubbers in new plants by requiring a percentage reduction in all SO2 emissions [19]. Very few coal-burning power plants have been affected by this requirement since only 179 of about 1,000 plants now in operation were built after 1978 [10].

We point out that the Clean Air Act does not address the problem of pollution emitted from "tall stacks" located in the midwestern U.S. These stacks are often over 1,000 feet high so that their emissions are swept into the upper atmosphere and blow over the areas in which they originate. These emissions generally fall back to earth over the eastern U.S. and Canada. The use of these stacks was encouraged by the U.S. Environmental Protection Agency as a cost-effective way for plants to meet local air quality standards [13].

Other highlights about legislation include the formation of the Interagency Task Force on Acid Rain by the U.S. Congress in 1980. Its mission was to conduct a 10 year study on this problem. In 1981, the "divisiveness" of the acid rain issue prevented Congress from renewing the Clean Air Act. Since that time, annual appropriation bills have been used to keep the act's regulatory provisions on the books.

3. THE DECISION MODEL

In this section, we first model the problem of predicting the acid rain policy for the U.S. using the AHP. Results are then presented using data based upon information compiled in the last year of the Reagan Administration.

3.1 Constructing the Hierarchy

To model this problem, we constructed the hierarchy that is shown in Figure 1. This hierarchy consists of 5 levels: Goal, Participants, Regions, Players, and Alternatives. The discussion that follows examines each level by starting with the goal and then proceeding top-down to the three policy alternatives.

Table 2. Clean Air Act Provisions

Name	Description
National Ambient Quality Standards	Enabled the EPA to set national limits on the emissions of what were considered the seven most damaging pollutants (including SO2 and NOx) from power plants and manufacturing facilities.
National Emissions Standards	To control sources of pollution.
State Implementation Plans	Allowed individual states to set their own limitations tailored to their needs.
New Source Performance Standards (NSPS)	Imposed technology-based emission control requirements on stationary sources under construction after 1971.
Statutory Mobile Source Control	Restricted auto emissions.
Prevention of Significant Deterioration	To prevent clean air areas from becoming polluted.
Limitations on New Emissions in Non-Attainment Areas	Control of pollution in areas where National Ambient Standards were not being met.

Table 3. New Source Performance Standards for Coal-Fired Plants

Pollutant	1971 Maximum Emissions	1978 Maximum Emissions
Sulfur Dioxide	No more than 1.2 pounds per million BTUs of any coal consumed.	No more than 1.2 pounds per mission BTUs of fuel consumed, plus 90 percent emissions reduction, or no more than 0.6 pounds per million BTUs of fuel consumed plus 70 percent emissions reduction.
Nitrogen Oxide	No more than 0.7 pounds per million BTUs of all anthracite, bituminous, and subbituminous coals consumed; 0.6 pounds for lignite.*	No more than 0.6 pounds per million BTUs of all anthracite, bituminous, subbituminous, and lignite coals consumed.
Particulate Emissions	No more than 0.1 pounds per million BTUs of fuel consumed.	No more than 0.03 pounds per million BTUs of fuel consumed.

* Anthracite, bituminous, and lignite are types of coal differentiated by inherent chemical composition.

Source: Congressional Budget Office [19]

3.1.1 Goal

The goal of our analysis is to answer the question: Which acid rain policy is the U.S. government likely to adopt in the next year given the current participants in the policy debate? Because about 97% of the SO_2 discharged into the atmosphere originates in plants built before the 1971 New Source Performance Standards (it is predicted that this level stays above 90% for the next 6 years [19]), older plants will be the primary targets for emission reduction strategies. Therefore, answers to the above question must focus on policy choices that have an impact on coal-burning plants already in operation or on plants under construction. In addition, the choice of a policy may also affect plants to be built in the future by specifying the required types of pollution controls or the types of coal that can be burned.

3.1.2 Participants

In examining the hierarchy, we see that the goal is decomposed into two major Participants that are considered the most influential in the decision process.

Figure 1. Hierarchy for Selecting Acid Rain Policy

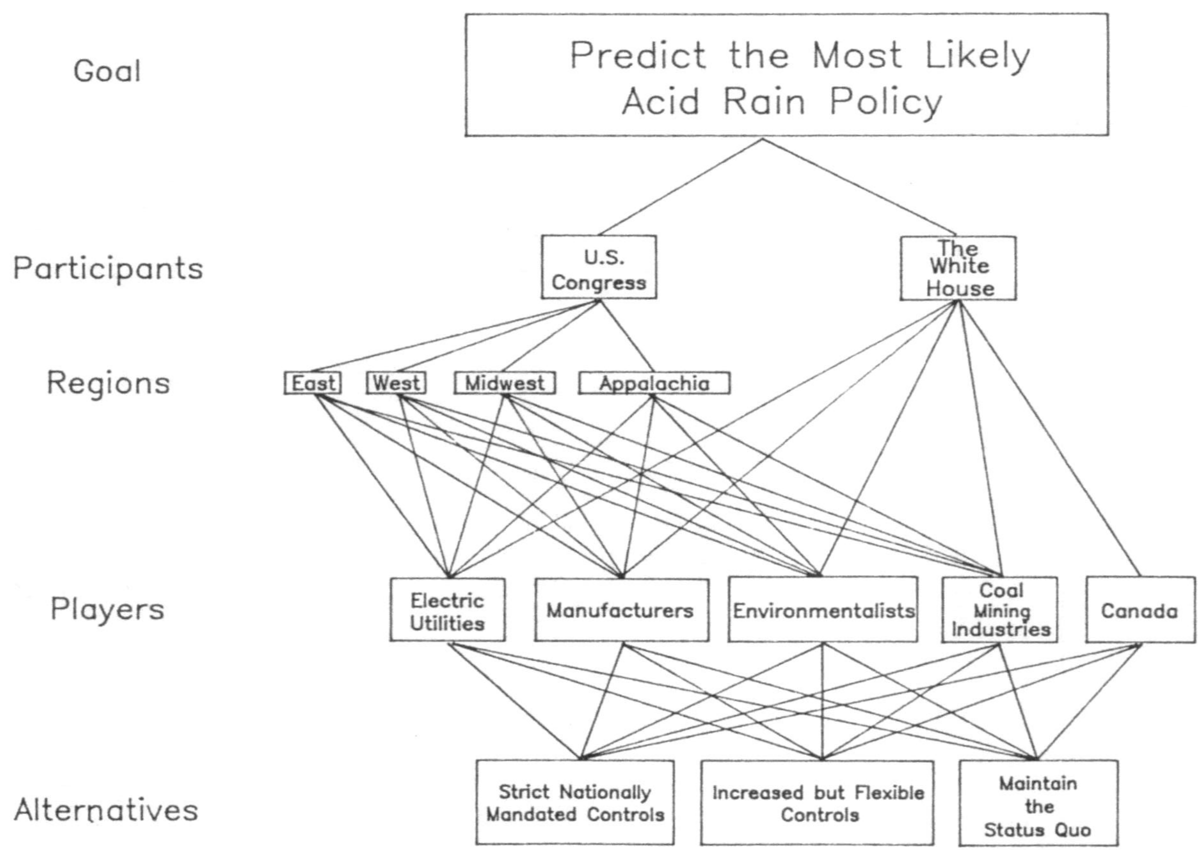

1. **U.S. Congress**--Members of Congress have significant input in selecting an acid rain policy for the U.S. In the last several years, the acid rain debate has become a "politically charged" issue that has been difficult to resolve. In both the House of Representatives and the Senate, the intense regional interests of several key, influential members (such as 1988 Senate Majority Leader Robert Byrd representing the coal miners of West Virginia) have served to create bipolar positions (i.e., either for strict controls or against strict controls) that have stalemated legislative action.

2. **The White House**--The power of the president in determining a U.S. policy should not be underestimated. The president appoints the Secretary of the Interior and the administrator of the Environmental Protection Agency--two individuals who have significant influence on the decision-making process. The president can also veto any congressional bill. In addition, he can appeal directly to the voting public. For example, from 1980 to 1988, the Reagan Administration consistently thwarted legislation by strongly voicing the opinion that not enough was known about the acid rain problem to justify regulatory action.

3.1.3 Regions

Since each member of the U.S. Congress is greatly influenced by the constituents that form his district, we break down the U.S. Congress into 4 key Regions (East, Appalachia, Mid West, and West) that have a significant stake in the outcome of the acid rain debate. Each region is represented by important Congressmen that can influence the choice of a policy.

1. **East**--This region includes New England, most of the Middle Atlantic states and the South Atlantic states (specifically, Alabama, Connecticut, Delaware, Florida, Georgia, Kentucky, Maine, Maryland, Massachusetts, Mississippi, New Hampshire, New Jersey, New York, North Carolina, Pennsylvania, Rhode Island, South Carolina, Tennessee, Vermont, and Virginia; the Appalachian sections of several states are listed in a separate region). Many states in the East are very interested in the choice of a policy since they have reported some damage due to acid rain. For example, a study by the U.S. Office of Technological Assessment reported in 1984 that 3,000 lakes and 23,000 miles of streams in the eastern U.S. had been acidified or were very close to acidification [20]. In the absence of stronger federal legislation, several states (e.g., Massachusetts and New York) have already implemented their own ambitious acid rain reduction plans [20].

There are several eastern politicians who are very active in debating this issue. Sen. Daniel Moynihan of New York favors moderate controls while Senators Robert Stafford of Vermont and George Mitchell of Maine support strict controls. Senator Mitchell has proposed Senate Bill S 321 that mandates a 12 million ton phased SO_2 emission reduction by the year 2000. This bill also sets lower emission limits than the current standards imposed by the NSPS and contains a provision that allows the federal government to control plants on an individual basis if state governments fail to alter pollution levels [1]. There is a 30 year shut-off rule for older plants not subject to the regulations of the Clean Air Act [5]. (Although not considered in this analysis, we point out that Senator Mitchell was elected Senate Majority Leader for the upcoming 101st Congress. This powerful position allows Mitchell to significantly influence legislation that will be introduced in 1989. For example, he determines which bills will be considered and voted upon by the entire Senate.)

2. **Appalachia**--This region contains the primary coal mining territories of the U.S.: West Virginia, western Pennsylvania, eastern Kentucky, and southwestern Virginia. Any type of acid rain legislation that might adversely affect the coal mining industry is opposed in this region. For example, regulations that require a shift to low sulfur coal would decrease the total amount of coal mined in this region. Since this area is already economically depressed, any action that might exacerbate the situation would be strongly opposed.

3. **Mid West**--This region contains heavy users of coal and a large concentration of manufacturers. It includes the states of Arkansas, Illinois, Indiana, Iowa, Kansas, Louisiana, Michigan, Minnesota, Missouri, Nebraska, Ohio, Oklahoma, Texas, and Wisconsin. This region also contains the largest producers of pollutants that are linked to acid rain.

The region's political responses to the acid rain debate are quite diverse. For example, Minnesota supports the "Mitchell Bill" (S 321), while Michigan and Ohio oppose it. In particular, Michigan and Ohio believe that such strict legislation would increase a company's cost of generating electric power, as well as increase energy prices for consumers. A strong opponent of additional acid rain legislation is Representative John Dingell of Michigan. He has consistently blocked attempts by the Congress to tighten clean air controls. For example, in 1984, he voted to remove acid rain controls from House of Representatives Bill HR 5314 [3]. Again in 1986, he opposed a plan to reduce emissions offered by Representative Henry Waxman of California, since he feared that this plan would have an adverse effect on the automobile industry [2]. We point out that as House Energy and Environment Committee Chairman, Representative Dingell has the power to curtail any proposed bills that were submitted for the committee's review.

In contrast, Minnesota has suffered considerable damage that is believed to be linked to acid rain. Representative Gerry Sikorski of Minnesota has sponsored HR 2666--a bill that proposes $SO2$ and NOx cutbacks and that would cost utilities between $4.5 and $5.3 million annually [1].

4. **West**--The states in the far west (i.e., Arizona, California, Colorado, Idaho, Montana, Nevada, New Mexico, North Dakota, Oregon, South Dakota, Utah, Washington, and Wyoming) appear to have suffered the least from acid rain. Less than 20% of the country's coal fired power plants are located in the West

and western coal averages 33% of the sulfur content found in eastern and midwestern coal [11].

Western politicians tend to favor strict controls. In 1984, Representative Henry Waxman of California sponsored a bill (HR 5314) that extended the Clean Air Act into 1990. A provision of this bill specifies an annual 14 ton reduction in SO2 emissions [7]. He later supported a phased reduction plan that would permit individual states to select their own reduction methods [2].

3.1.4 Players

Each of the regions is further decomposed into four key Players (Electric Utilities, Manufacturers, Environmentalists, Coal Mining Industries) that are very influential in the decision-making process.

1. **Electric Utilities**--Power companies use coal to generate over one half of the electrical needs of this country; they also produce between 60% to 70% of the SO2 emissions found in the atmosphere [19]. Given a policy of strict controls, utilities feel that the cost of generating electricity would increase substantially, thus increasing consumer costs. If wet-scrubbers were mandated by law, one source estimates that the consumer's price for electricity would increase 11.3% over 5 years [11].

The utilities have adopted the position that (a) current laws have effectively reduced dangerous emissions, (b) further research is needed to establish that fossil fuel emissions are linked to environmental damage [11], and (c) clean coal technologies cannot be pursued at the same time that costly pollution controls are being installed and maintained [6]. Organizations such as the Edison Electric Institute and the American Public Power Association use their influence to steer the U.S. government away from a policy of strict controls.

2. **Manufacturers**--This player represents a diverse group of industries (e.g., aluminum, automobile, chemical, and steel manufacturers in the Mid West [11]) that use coal as an input to a production process or that rely on coal-fired plants for electricity. They fear that the adoption of strict pollution policies may lead to higher manufacturing costs that are tied to the rising cost of electricity. Costs will also increase if industries are required to install pollution control equipment such as wet-scrubbers. Key lobbying groups include the National Association of Manufacturers and the Electric Consumers Resource Council (ELCON). ELCON claims not to have found any scientific evidence to indicate that a reduction in SO2 emissions will generate benefits equal to the cost of implementing a reduction policy. Therefore, it favors no change from the status quo.

3. **Environmentalists**--Groups such as The Sierra Club, The National Clean Air Coalition, The Natural Resources Defense Fund, The National Wildlife Federation, and The Environmental Defense Fund are united in their campaign to quickly stop the emission of substances linked to acid rain. In past years, these groups have exerted a strong influence on the policy debate through lobbying,

public education efforts, and legal actions. For example, The Natural Resources Defense Fund has successfully confronted General Motors, National Steel, and Monsanto in court over air pollution issues [11]. The Sierra Club employs a full-time staff of attorneys to prosecute environmental cases, while The Environmental Defense Fund supports a powerful lobby known as the National Acid Rain Initiative Ballot [11].

4. **Coal Mining Industries**--There are an estimated 56,000 jobs in the U.S. that are linked to high sulfur coal mining [19]. The type of policy that the government adopts has an immediate effect on this industry. For example, a policy that would reduce SO_2 emissions by 10 million tons would eliminate about 15,000 to 22,000 jobs [19]. Alternatives that encourage the use of low sulfur coal would shift the major coal supply points from Appalachia and the Mid West to the West where low sulfur coal is in good supply. This industry is represented by groups such as The National Coal Association, The Ohio Mining and Reclamation Association, and The Alliance for Clean Energy.

The four key players that we have just described and an important fifth player (i.e., Canada) are included under the White House node of the hierarchy.

5. **Canada**--The Canadian Department of Agriculture estimates that on average 50% of the acid rain that occurs in Canada originates in the U.S. [17]. In some regions (such as eastern Canada) this figure may be as high as 70% [14]. The country has strong laws governing acid rain and for many years has pressured the U.S. to enact stricter legislation. Canada has successfully created cooperative regulatory programs with several individual U.S. states including Connecticut, Maine, Massachusetts, New Hampshire, and Vermont [18]. However, it has also been accused of pursuing a reduction in U.S. emissions in order to gain economic benefits. For example, in 1984, the president of The Ohio Mining and Reclamation Association claimed that Canada wanted a 50% decrease in U.S. emissions (recall that installing control methods like wet-scrubbers might decrease the operational efficiency of plants and, hence, reduce the amount of electricity they produce) so that new plants with excess capacity located in Ontario could "pick up the electric slack" due to the potential decrease in the output of Ohio-based plants [11].

3.1.5 Alternatives

At the bottom level of the hierarchy are shown the three policy options: Strict Nationally Mandated Controls, Increased but Flexible Controls, and Maintain the Status Quo. In the discussion that follows, we briefly summarize each alternative.

1. **Strict Nationally Mandated Controls**--This alternative would require manufacturers and utilities to dramatically reduce emissions from existing facilities to levels below NSPS standards over the next 5 to 10 years. Many western states would be exempt from controls while coal-fired power plants and industrial facilities in the remaining states would be required to install

wet-scrubbers. This option could take the form of legislation that is similar to the Mitchell Bill.

2. **Increased but Flexible Controls**--With this option, plants and manufacturers would be allowed flexibility in meeting tougher emission standards that are similar to NSPS reduction requirements for existing plants. Switching to low sulfur coal would satisfy most short-term requirements (over the next 2 to 5 years) and, at the same time, allow coal users to develop, test, and market new clean coal technologies.

3. **Maintain the Status Quo**--The current version of the Clean Air Act, as well as the regulations of individual states, would be relied upon to control emissions. Research into the link between acid rain and environmental damage would continue to be funded.

3.2 Executing the Model

In the summer of 1988 (while the Reagan Administration was still in power), we compiled information relating to the various levels of the hierarchy and used this data to generate entries for each of the pairwise comparison matrices. The Expert Choice software system was used to facilitate our analysis. As mentioned earlier, our model and judgments reflect the viewpoint of an electric utility involved in a yearly planning process. Since 1988 was an election year, we also modified our judgments to reflect the choice of presidential candidates. The results are summarized in Table 4.

Table 4. Final Set of Weights

	Administration		
Alternative	Reagan	Bush	Dukakis
Strict Controls	.262	.264	.343
Increased but Flexible Controls	.276	.310	.314
Maintain Status Quo	.462	.426	.342

From Table 4, we see that the Reagan Administration is strongly in favor of maintaining the status quo. Throughout his tenure, Ronald Reagan stated that more research was required before the government could take action to control acid rain. It was only in his second term that he "officially" recognized that a problem existed and committed $5.5 billion to support research. Although, during his campaign, George Bush pledged to clean up the environment, we believe that he will work only to maintain current laws and regulations that limit emissions. However, if Michael Dukakis had been elected president, then the U.S. policy on acid rain would have moved much closer to imposing strict national controls.

4. CONCLUSIONS

In this paper, we have provided decision makers in the electric utility industry, as well as in the automobile, steel, and aluminum industries with an AHP-based template that can be used to help predict which acid rain policy will be selected for the United States. We envision our model being used on a yearly basis to help make planning decisions. It is a flexible model that can be easily altered to reflect either changes in the country's leadership or shifting views on acid rain.

The model and results that we present in this paper have been carefully reviewed by the Maryland-based Baltimore Gas and Electric Company (BG & E). This company serves nearly one million customers and uses a wide variety of coal types to generate over 40% of its electricity (nationally, coal was used to generate nearly 56% of the total amount of electricity in the U.S. during 1987 [16]). This percentage will increase when a new coal-fired plant is finished in the early 1990s. The decision process that resulted in BG & E selecting a conventional coal-fired plant is reported in [12]. We point out that as part of this analysis, the decision makers needed to assess future clean-air regulations. In order to comply with environmental regulations, BG & E and other utilities have increased capital expenditures.

Though specific costs regarding BG & E cannot be made available, estimates of electric utility costs nationally can be reported. According to EPA data, the electric utility industry has spent or will spend about $10 billion annually on environmental control devices between 1981 and 1990 [4]. As of December 31, 1986, privately owned electric utilities reported that $36.3 billion of electric plant-in-service costs, representing 11.2% of the total, is environmental protection related [4]. In 1986, $2.2 billion of operating expenses were also related to environmental protection devices or programs [4]. These figures do not include the cost of purchasing low sulfur coal.

The electric utility industry is concerned that the acid rain policy options currently under review by the U.S. government might substantially increase capital, operation, and maintenance costs for environmental control equipment in coal-fired power plants that are already on line. For example, according to EPA estimates in a report prepared for the Business Roundtable in March 1988, legislation similar to the Mitchell Bill would increase capital costs for SO2 and NOx pollution control by about $9 billion annually [9].

BG & E is considering using the results of an AHP-based acid rain model (similar to the one constructed in this paper) as input into their strategic planning process. This process includes forecasting both long-term electric power plant needs and future fuel costs.

5. REFERENCES

1. Anonymous, "Acid Rain and Clean Air Legislation Still Unfinished," **Public Utilities Fortnightly, 120, No. 3**, 25-26 (1987).

2. Anonymous, "Acid Rain Bill Stalls," in **Congressional Quarterly Almanac**, U.S. Government Printing Office, Washington, D.C. (1986).

3. Anonymous, "Clean Air Bill Stalled by Acid Rain Dispute," in **Congressional Quarterly Almanac**, U.S. Government Printing Office, Washington, D.C. (1984).

4. Anonymous, "EIA Financial Statistics of Selected Electric Utilities: 1986," U.S. Government Printing Office, Washington, D.C. (1987).

5. Anonymous, "Senate Environment Moves on Clean Air; Compromises More Difficult Than Original," **Coal Week, 13, No. 43**, 1-3 (1987).

6. Anonymous, "SO2 Emissions Drop Again; Acid-Rain Bill Denounced," **Electric Light and Power Monthly, 66, No. 9**, 10 (1988).

7. Anonymous, "The Acid Rain Controversy," **The Congressional Digest, 64, No. 2**, 33-64 (1985).

8. B. Burke, "A Look at Air Quality Trends," **EPA Journal, 13, No. 8**, 25-27 (1987).

9. R.M. Dowd & Company, "Analysis and Impact of S 1894: The Clean Air Standards Attainment Act of 1987," Prepared for the Business Roundtable, B2-5 (1988).

10. Edison Electric Institute, **Power Directory** (1986).

11. A. Kahan, **Acid Rain: Reign of Controversy**, Fulcrum Incorporated, Golden, Colorado (1986).

12. R. Keeney, J. Lathrop, and A. Sicherman, "An Analysis of Baltimore Gas and Electric Company's Technology Choice," **Operations Research, 34**, 18-39 (1986).

13. J. Luoma, **Troubled Waters, Troubled Skies: The Story of Acid Rain**, The Viking Press, New York, New York (1984).

14. T. McMillan, "Why Canadians Worry About Acid Rain," **EPA Journal, 12, No. 5**, 8-10 (1986).

15. V. Mohnen, "The Challenge of Acid Rain," **Scientific American, 259, No. 2**, 30-38 (1988).

16. North American Electric Reliability Council, "1988 Electric Supply and Demand for 1988-1997," 38 (1988).

17. E. Somers, "Environmental Hazards Show No Respect for National Boundaries," **Environment, 29, No. 5**, 7-9 (1985).

18. R. Thompson, "Acid Rain: Canada's Push for U.S. Action," in **Earth's Threatened Resources**, U.S. Government Printing Office, Washington, D.C. (1986).

19. United States Congress: Congressional Budget Office, **Curbing Acid Rain: Cost, Budget, and Coal Market Effects**, U.S. Government Printing Office, Washington, D.C. (1986).

20. M. Weisskopf, "Impatient with U.S., States Acting on Acid Rain," **The Washington Post, 1988, No. 145**, A7 (1988).

DECISION SUPPORT FOR WAR GAMES

Robert J. Might
War Gaming and Simulation Center
Institute for National Strategic Studies
National Defense University
Ft. L.J. McNair
Washington, D.C. 20319

William D. Daniel Jr.
CACI Products Company
Arlington, Virginia 22209

ABSTRACT

This paper discusses a decision support tool that can help players of a war game simulation decide on the most efficient use of military forces and limited airlift and sealift required to move the military forces. The Analytic Hierarchy Process is employed to develop military unit values that are used as coefficients in a linear program that assigns forces to different regions of the world during crisis decision exercises.

1. INTRODUCTION

Crisis decision exercises (CDE's) are a type of political-military simulation conducted with the aid of a computer model in which players at the National Defense University try to decide on the best relocation of forces to desired areas of operation. In determining the best strategy, players need to focus on four issues: (1) U.S. global interests, (2) adversary intentions, (3) adversary capabilities, and (4) the level of risk associated with the choice of different options. Players make decisions at the National Command Authority Level (i.e., at the level of the President and the Secretary of Defense). At this high level, a decision maker does not deal with detailed alternative deployment schemes for U.S. forces but rather considers the overall deployment of forces.

In a CDE, a player needs feedback on what are feasible military courses of action and what are the limiting factors. A computer model that can support this type of deployment decision making should have the following characteristics:

1. It must provide realistic information;
2. It must limit the deployments to those that are feasible in the time allocated by the decision maker (for example, the model must account for the availability of airlift and sealift);
3. It must consider the sustenance of the units once they reach their deployment region;
4. It must offer the player the option to select the units to be deployed in each region of the world and it must

be capable of making that decision with notional or actual units (e.g., one airborne division or the 82nd Airborne);

5. The decisions related to the selection of forces must consider important military factors such as the terrain, road conditions, extreme weather conditions, and the most likely enemy actions in that region;

6. The aid should provide time-line presentation of what units are at their deployment region, what portion of transportation assets (both sea and air) is in use or obligated at any point during deployment, what follow-on forces are available, and what their deployment and sustenance schedules and costs should be.

This list is not exhaustive. However, adding requirements could cause the decision model to become very complex. The need for model accuracy has to be balanced against ease of use and fast run times.

Players need to discover the impacts and limitations of major decisions during a crisis. They do not want to dwell on the details of unit deployment. For example, during an exercise, a player will be interested in how small changes in U.S. regional priorities change worldwide response requirements. Similarly, if a player wants to know how the U.S. capability to respond varies with changes in the enemy's intentions, there should be a way to rapidly assess the impact of different estimates. The decision support tool that is described in the next section can provide first-order impacts of alternative decisions or estimates.

2. MODEL DESIGN

To use a decision aid effectively, a player should only have to consider a few parameters: (1) regional priorities--review all regions under consideration for deployment of forces, assess each region's political, economic, and military importance to the U.S., and rank the regions based on the situation, (2) enemy intentions --establish the purpose or objective of enemy actions globally and in each region, (3) enemy capabilities, and (4) acceptable risk in each region--since the U.S. cannot be strong everywhere, all regions cannot be provided with sufficient forces for a low risk of loss to enemy action; in what regions can the U.S. then afford a medium to high risk of loss? Players would input these four parameters at key decision points in the simulation. Additional parameters, such as the allocation of airlift and sealift to each region, can be considered when the overall ranking of regions is generated.

We propose to model the deployment decision problem and to take into account the four parameters mentioned earlier by using a linear program that has the characteristics of a classic transportation problem. The objective function incorporates the perceived "value" of each combat unit in a particular region for the situation that the player estimates will exist in that region.

In general terms, the objective function is given by:

$$\text{Maximize } \Sigma_{time}\Sigma_{regions}\Sigma_{units}U^{i}_{vtr}RP_{t}$$

where U^{i}_{vtr} is the value of unit i in time period t in region r and RP_t is the regional priority in time period t. The objective is to maximize military value across all regions and will account for the units in a region prior to deployment, the units deployed from a region, and the units that are re-deployed to a region. There are four major sets of constraints that are described below.

1. **Lift constraints**--These restrict the amount of air and sea lift that is available for each region to a predetermined level or a player-specified percentage of the total lift.
2. **Unit deployment constraints**--These allow players to prohibit movement of units from one region to another (e.g., no units will be deployed from Europe to another region).
3. **Sustainability constraints**--These constraints require units to be resupplied for all of the time periods in which they are deployed. Players must take into account those levels of forces, materials, and consumables necessary to support a military effort.
4. **Risk constraints**--The player must decide on the level of acceptable risk in each region. Again, risk can be defined as the degree of danger or uncertainty that a player is willing to accept that (a) his intentions will be foiled by the enemy, and that (b) he may or may not be able to foil enemy intentions. The choices range from high risk (i.e., friendly forces are easily defeated) to low risk (i.e., enemy forces are easily defeated). Among the model's parameters, is a friendly-to-enemy force ratio that is specified for each level of risk. When this ratio matches the degree of risk specified by the player, this set of constraints will stop the deployment of friendly forces to that region. Without these constraints, most available units would be sent to the regions with the highest priorities and actual force requirements for a region would be exceeded. This would result in low priority regions receiving no reinforcement due to lack of forces.

It is important to note that a player's estimate of the enemy's intentions has a significant impact on the set of risk constraints. For example, if a player chooses a low risk approach in a region where he believes that the enemy will attack, then the forces that are deployed to this region will be wasted assets should the enemy really take a defensive posture.

The computer implementation of our LP model consists of three separate modules: (1) a pre-processor that develops the necessary coefficients for the LP, (2) a standard mathematical programming package that solves the LP, and (3) a post-processor that converts the output from the LP into an appropriate format that is easily understood by the players. The remainder of this paper focuses on the use of the Analytic Hierarchy Process to develop the LP objective function coefficients (i.e., the values of U^1_{vtr}).

A typical formulation will contain approximately 200,000 decision variables. For a deployment to Southwest Asia, a small problem of about 25,000 decision variables can easily be run on a VAX 11/785. An interesting global deployment could easily exceed 500,000 decision variables. Informal discussions held recently with Military Airlift Command (MAC) Headquarters may lead to running the LP on their AT&T KORBX computer with the Karmarkar solution procedure. Similarly sized linear programs have been solved parametrically on an IBM mainframe computer with the MPS III mathematical programming package.

3. DETERMINING MILITARY UNIT VALUES

The most difficult coefficients to specify in the LP are those related to combat unit selection for deployment in each region. For example, Table 1 gives a partial listing of those factors that can impact a combat unit's effectiveness.

Table 1. Factors Affecting Combat Unit Selection

Factor	Components
Environmental	Terrain Weather Road conditions Obstacles (e.g., rivers)
Situational	Friendly unit posture (such as an attack posture or defend posture) Enemy posture Degree of preparation Level of attacker's surprise
Unit & Enemy Strength	Equipment and number of people Supplies on hand Morale and cohesion Leadership capability Recent combat Level of training

Trying to combine the large amount of information presented in Table 1 in order to produce a "military value" for each unit is a difficult and time-consuming task. We will use the AHP to develop values for a unit in each region of the world for each situation in which it can be deployed. There are significant advantages in using the AHP.

1. Many of the issues that must be considered when developing the military values are subjective in nature and cannot be assessed through standard analytic tools. However, these important subjective criteria can be included when the AHP is used to establish preferences.

2. Many military units do not have lethal missions. An example is the Airborne Warning and Control Systems (AWACS). Thus it is very difficult to include these units in standard force effectiveness measurements. Yet these units are very important to the combat outcome and they also require the use of airlift and sealift to deploy. The AHP allows the decision maker to assess and compare the importance of these units in the overall deployment strategy.

3. There is no force measuring technique that allows analysts to place U.S. Navy units on an "effectiveness scale" similar to Army and Air Force units. For example, a U.S. armored division is normally used as a base ground force unit and assigned an effectiveness score of 1.0. There is agreement among military experts that a mechanized U.S. infantry division should score about 0.97 and an armored brigade should score about 0.33 (the brigade is about one-third the size of an armored division). There is no similar ranking scheme that can measure the effectiveness of individual ships or clusters of ships. The AHP can help convert subjective preferences regarding naval units into a numerical effectiveness measure.

4. The AHP can determine military unit values in a formal and structured manner without requiring the collection and analysis of large amounts of data on military units, terrain, weather, and other factors.

In order to use the AHP in the deployment decision model, the simulation designer must: (1) define global regions for force employment (employment is the placement and use of units in a combat zone to accomplish a military purpose), (2) establish force postures (i.e., the physical and geographical configuration of a unit required to conduct a tactical operation), and (3) decide upon ground, air, and naval force aggregations for measurement.

In the first step, the designer must define the regions of the world that are of interest. This provides for a very flexible tool that can be used at different levels of decision making. For example, if the exercise is a National Command Authority level game, the regions of interest might correspond to the established U.S. unified command areas (this is a command with a broad continuing mission under a single commander and composed of two or more services). However, if the U.S. commander of forces in Europe is interested in deployment within his theater of

Table 2. Seven Military Conditions

Military Condition	Definition
Deliberate Attack	A type of offensive action characterized by preplanned coordinated employment of firepower and maneuver to close with and destroy or capture the enemy.
Hasty Attack	An offensive operation launched with the forces at hand and with minimum preparation to destroy the enemy before he is able to concentrate or establish a defense.
Deliberate Defense	A defense normally organized when out of contact with the enemy or when contact with the enemy is not imminent and time for organization is available. It normally includes an extensive fortified zone.
Hasty Defense	A defense normally organized while in contact with the enemy or when contact is imminent and time available for organization is limited. It is characterized by improvement of the natural defensive strength of the occupied terrain.
Delay	A military operation in which a force under enemy pressure trades space for time by slowing down enemy momentum and inflicting maximum damage on the enemy without becoming decisively engaged.
Feint	The act of drawing the attention and focus of an enemy from the point of principal operation.
Withdraw	A maneuver whereby a friendly force disengages from an enemy force in accordance with the will of the commander.

operation, the regions could correspond to the geographical boundaries of his subordinate Corps commanders.

The current model database considers deployments within nineteen global regions that are generally subsets of established U.S. military unified command areas. The regions are further divided into countries, sectors, and borders. This organization allows the decision model to consider deployment problems at several levels: (a) all nineteen world regions, (b) a group of regions, e.g., northern, central, and southern Europe, (c) a single region, (d) a single country within a region, and (e) a regional border.

After the regions are defined, the AHP is used to establish a preference for each type of military unit in each of seven military conditions that are displayed in Table 2. It is important to note that actual force preferences are made from the point of view of the player who is conducting the operation. The preferences usually focus more on regional considerations than on an opponent's actions or capabilities.

We now construct the hierarchy for determining military unit values. Since almost all sizeable military operations will consist of at least two of the three military services, i.e., ground, air, and naval forces, we use these services as the top level in our hierarchy. However, every unit category within these three components will not be equally useful in each region and in each posture. Therefore, it is necessary to identify a specific type of unit in each component that can serve as a **linch-pin category**. Figure 1 illustrates this concept. At the top of this figure are the three components: Ground, Air, and Naval. Three linch-pin categories are then specified: U.S. Armored Division, USAF F-16 Wing (trained and equipped for air-to-ground operations), and U.S. conventional powered carrier battle group. Below these are listed all of the other units within a component. (Although not shown in Figure 1, we could also list tactical unit types. For example, a C-130 transport squadron might appear under the air component.) We then begin our pairwise comparisons of the units within a component against the unit specified in the linch-pin category.

It is important to realize why forces must be tied to a specific linch-pin category. For example, if the commander in Northern Europe was in a hasty attack posture, a U.S. armored division would be of significantly less value than an air assault division since the terrain is not well suited to armored vehicles. The commanders in each region must have the same "capability" in mind when they compare the importance of ground, air, and naval forces. This approach reduces the chances of significant differences due to the player's experiences and biases. The criteria that a player should use when making comparisons are listed in Table 3.

Figure 2 displays the weights that result from a comparison of the unit categories in Northern Europe in a hasty attack posture. In making these judgments, we note that air units must be treated differently than the ground and naval units. It is

Figure 1. Components and Categories

COMPONENTS

Ground	Air	Naval

LINCH-PIN CATEGORIES

U.S. Armored Division	USAF F-16 Wing	U.S. Conventional Powered Carrier Battle Group

UNIT CATEGORIES

Armored Division	Air Superiority	Amphibious Task Force
Infantry Division	Airborne Warning and Control System	Battleship Surface Action Group
Air Assault Division	Deep Air-to-Ground	Cruisers (Nuclear) Cruisers (Conventional)
Airborne Division	Electronic Warfare/Suppression of Enemy Air Defenses	Carrier Battle Group (Conventional Power)
Light Division	Shallow Air-to-Ground	Carrier Battle Group (Nuclear Power)
Mechanized Division	Tactical Airlift	Destroyers and Frigates (All Classes)
Motorized Division	Tactical Reconnaissance	Attack Submarines (Nuclear Power)

Table 3. Criteria for AHP Selections

For a given region and posture consider the following criteria:

---

1 **Unit Firepower**
Consider firepower available from unit-held weapons only. Do not consider the additional capability that might be provided by attachment of other units or by normal supporting or reinforcing forces.

2 **Terrain and Weather**
What effects inherent in the terrain and weather for the region must be overcome or turned to a friendly advantage? In each case, how well does the unit being considered perform in these circumstances?

3 **Unit Tactical Mobility**
For this region and posture, does the unit being considered possess a high (or low) degree of tactical mobility? (Tactical mobility is the degree to which a military unit can move rapidly and frequently within a theater or on the battlefield. For example, in the jungle, an airborne division has high tactical mobility while an armored division has low mobility.) The mobility differential between friendly and probable enemy forces should also be considered.

4 **Enemy Sophistication and Quality**
Is the expected opposition a well-trained, high-technology supported force or a recently activated, third-world reserve force with limited support capability?

5 **Allied Capability in Area**
Are there allied forces available to operate with the U.S. force being considered? Do these forces eliminate or reduce any known shortcomings in U.S. capability?

6 **Degree of Air and Sea Control**
To what degree does the U.S have freedom of action in the air and sea approaches to the region and within the region. The lack of sea and air superiority (or parity) will have a significant impact on the military value of units in the region.

Figure 2. Military Unit Values from the AHP

COMPONENTS

Ground	Air	Naval

LINCH-PIN CATEGORIES

U.S. Armored Division	USAF F-16 Wing	U.S. Conventional Powered Carrier Battle Group
.429	**.429**	**.143**

UNIT CATEGORIES

Armored Division **.015**	9 AWACS **.128**	Amphibious Task Force **.014**
Infantry Division **.063**	3 F-15 Squadrons **.091**	Battleship Surface Action Group **.021**
Air Assault Division **.189**	3 F-16 Squadrons **.086**	Cruisers (Nuclear) Cruisers (Conventional) **.013**
Airborne Division **.055**	3 RF-4 Squadrons **.041**	Carrier Battle Group (Conventional Power) **.033**
Light Division **.055**	3 C-130 Squadrons **.041**	Carrier Battle Group (Nuclear Power) **.047**
Mechanized Division **.020**	3 F-15E Squadrons **.022**	Destroyers and Frigates (All Classes) **.009**
Motorized Division **.031**	3 F-4G Squadrons **.020**	Attack Submarines (Nuclear Power) **.005**

possible for a player to picture a typical mechanized infantry division or a nuclear cruiser. Even though specific divisions and individual ships can have different capabilities, the overall characteristics are very similar. However, the aircraft that can be placed in the different units can have very dissimilar characteristics and therefore it is essential for a player making comparisons to know which aircraft are representative of the unit type. This requires a comparison of the different types of aircraft among units.

Since many of the regions of interest to U.S. policy makers and military planners have allied military units that would fight alongside the U.S. units, our model must account for this added strength. To accomplish this, we assign values to allied units that are relative to U.S. units of the same type. For example, a West German armored division could be assigned a relative score of 0.87 based on the 1.0 score of a U.S. armored division. This differential would be calculated on the basis of factors such as equipment, training, and morale.

Finally, we point out that maintaining a current database of military values is very tedious and time-consuming when there is a single expert for a region. It becomes quite difficult when several expert comparisons must be combined.

4. CONCLUSIONS

We have shown that the AHP can use the preferences of experts to determine military unit values in a formalized and structured manner without requiring the collection and analysis of large amounts of information on factors such as units, terrain, and weather. The values can then be used as part of a fast-running LP deployment model in a simulation framework to efficiently determine the best allocation of military forces to different regions of the world. We point out that data on military unit values can potentially be useful in other decision-making situations. For example, if an arms-control negotiator must assess a proposal and has very little time for analysis, a quick summation of the military values of the forces in opposing postures would provide significant insight into the proposal.

ASSESSMENT OF SECURITY AT FACILITIES THAT PRODUCE NUCLEAR WEAPONS

John G. Vlahakis
Office of Civilian Radioactive Waste Management
U.S. Department of Energy
1000 Independence Avenue, S.W.
Washington, D.C. 20585

William R. Partridge
Surveys and Investigations Staff
U.S. House Appropriations Committee
Room 1719, HHS Building
3rd & Independence Avenue, S.W.
Washington, D.C. 20201

ABSTRACT

The AHP has been applied to the highly complex problem of determining the adequacy of security at selected U.S. Department of Energy facilities that produce nuclear weapons. Use of the AHP enabled an Inspector General's team to determine the relative importance of specific countermeasures at these facilities. Thus, the team was able to weight and rank recommendations for corrective action where deficiencies were found to exist. The hierarchy, which included a variety of factors, provided the Department of Energy with a flexible tool for assessment, planning, and allocation of resources for enhanced security.

1. INTRODUCTION

Expenditures on security systems, in general, have risen sharply in recent years. This is due to increased concern about threats and to the availability of new technologies. From the point of view of the U.S. Department of Energy (DOE), nuclear weapons security refers to the monitoring and protection of certain valuable materials as they are acquired, produced, and stored. The DOE is responsible for the design, development, production, and testing of nuclear weapons for the Department of Defense. Weapons-related activities account for approximately two-thirds of the DOE budget.

The problem facing assessors of security involves a multi-criteria evaluation of the detailed elements of the security at nuclear weapons production facilities. The more systematic this evaluation becomes, the easier it is to follow (and accept) an assessor's line of reasoning. This paper describes an attempt to systematize the evaluation process that began in 1984 when the authors (then working in the Office of the Inspector General at the Department of Energy) were asked by the Inspector General (IG) to perform a comprehensive assessment of the security at nuclear facilities. Certain details of the data and project have been altered or deleted so that this application can be published in unclassified form.

The nuclear weapons production system consists of numerous DOE facilities located throughout the United States. Some facilities have only small amounts of material and equipment, making them attractive targets for theft or sabotage. Others have large quantities of valuable material in a form that makes theft or sabotage less likely, because the material would be difficult to handle and use. Still other facilities have assembled components and completed weapons that might be extremely attractive to an adversary. In addition, there is a transportation system connecting facilities, DOE, and military installations. Some form of security is required for each of these operations, but the type and level of security varies considerably depending on the specific operation.

2. SPECIFIC MOTIVATION

Security measures are instituted to respond to threats, whether real or perceived. Ideally, an assessment of effectiveness should be made in terms of meeting these threats. Previous security reviews by the IG had revealed the fact that few attempts had been made by security managers to relate countermeasures to postulated threats. Most budgetary justifications for security enhancement were made by experts in security systems who apparently were not required to relate their recommendations to specific types of threats.

One reason for this is clear. Most security personnel were not experts on potential types of adversarial action. Another reason is that adversarial action against nuclear facilities had been practically nonexistent. There was simply no experience upon which frequency-based statements of likelihood could be developed. The only alternative was to develop threat assumptions derived from analogous experiences, feared consequences, and the like. This had been done at DOE, but only in the most general terms.

DOE officials issued a classified memorandum in 1983 which described five categories of adversary and broad strategies these adversaries might be expected to use for theft or sabotage. This generic threat guidance did not state which of these adversarial types or strategies were more important than others. Site managers were directed to take these threats into account in designing security systems to protect against acts of sabotage and theft or the diversion of special nuclear materials (SNM). This threat guidance was considered too vague to serve as a baseline for the IG's assessment.

Given our charge, we sought to weight and rank the importance of security subsystems on a department-wide basis, rather than on a site basis, before checking the subsystems for adequacy. If an identical weakness is found in the same security subsystem at two sites, it is likely that corrective action at one site may be more important than similar action at the other. We needed to obtain expert advice, determine where the weaknesses were, and deal with priorities. We set out to develop a perspective from which we could advise DOE managers where money for security enhancement would do the most good.

3. METHODOLOGY

It was decided that the project would be conducted in two phases. The ultimate objective was to assess security at 22 sites. The primary purpose of Phase 1 was to develop the methodology and criteria for assessment. In particular, Phase 1 was designed to include "pilot" assessments of four sites and the intersite transportation system. In response to real security concerns, we selected four sites that were most sensitive from a security standpoint. DOE managers planned to initiate corrective action immediately if required. A parallel effort during Phase 1 was to establish criteria for assessing the quality of individual security components. Eight inspectors were assigned to this project for a period of one year at a cost of $1 million. Most of the effort was directed at criteria development and pilot assessments.

We decided to use the AHP to assess security and the hierarchy was developed during a meeting at which we assembled numerous experts including Professor Thomas Saaty. The hierarchy was structured to address target attractiveness (consequence areas, inherent material attractiveness), threat scenarios (adversary scenarios, types, tactics), and countermeasures (functions and subsystems). Each level of the hierarchy is depicted in Figure 1.

For the most part, detailed discussions were held by two separate groups. One group focused on threats, while the other concentrated on countermeasures. The ability to have different experts make judgments about different parts of the hierarchy was one of the advantages of the AHP model.

We elicited judgments from the experts on the relative importance of elements at each level, asking the following questions:

Level 2. If the security of a nuclear weapon facility is compromised, which consequence area is more important with respect to the public interest?

Level 3. Given a consequence area, which strategy impacts that consequence area more severely?

Level 4. Given a strategy, which adversary type is more likely to have the capability and interest to carry it out?

Level 5. Given an adversary type, which target type is more likely to be selected?

Level 6. Given a target type, which is the more credible and effective tactic to be chosen?

Figure 1. Schematic View of Hierarchy

Level	Name	Examples of Elements
1	Goal	Importance of Security System Components
2	Consequence Area	National Security, Public Health and Safety, Political and Social
3	Adversary Strategy	Theft of Weapon, Detonation of Weapon
4	Adversary Types	Terrorist, Disgruntled Employee
5	Target Types	Weapons or Components, Chemical Compound
6	Adversary Tactics	External Assault, Deceptive Action
7	Countermeasure Functions	Deter Insider, Delay Access
8	Countermeasure Subsystems	Passive, Protective Force
9	Countermeasure Components	Fences, Guards, Alarms, Background Investigations

Level 7. Given adversary tactics, which countermeasure functions are most effective in reducing the likelihood of a successful attack?

Level 8. Given a countermeasure function, which subsystems are most effective?

Level 9. Given a countermeasure subsystem, which components are most reasonable?

Differences in importance were judged by the standard AHP comparison procedure. We used Expert Choice throughout the project to facilitate this procedure, display results conveniently to participants, and perform sensitivity analysis. The consensus of the groups was calculated using the geometric mean of individual judgments. Particular care was taken to phrase questions asked at each level so that relative importance was judged rather than probability. Clearly, security of nuclear weapons relates to problems of high consequence but low probability.

4. DETAILS OF THE HIERARCHY

The goal node represents the first level of the hierarchy. The second level of the hierarchy consists of areas of potential consequences of adversarial action, such as:

- o National Security,
- o Public Health and Safety,
- o Political and Social.

Originally, there were six consequence areas. In the judgment of our experts, three consequences were considered much less important than the others and they were dropped to simplify the hierarchy. These were Occupational Health and Safety, Economic, and Environmental. The last area, Environmental, was considered to be encompassed by Public Health and Safety.

The third level of the hierarchy consists of adversarial strategies, such as:

- o Theft of Weapon,
- o Detonation of Weapon,
- o Theft of SNM,
- o Radiological Sabotage.

One of the ground rules of this project was that theft of information and all forms of espionage would not be considered. "Theft of Weapon" means theft of a nuclear device ready for testing or delivery to the military. "Detonation of Weapon" means detonation without removal of the device from the site. "Theft of SNM" means theft of weapons components or Special Nuclear Materials. "Radiological Sabotage" means action against a site that has the effect of dispersing radiological material at the site and its surrounding area.

Five adversarial types were defined at the fourth level of the hierarchy:

- o Terrorist (with insider assistance),
- o Disgruntled Employee,
- o Criminal (insider or outsider),
- o Foreign Agent,
- o Anti-Nuclear Extremist.

Invasion by a foreign power, such as a large force conducting a commando-type raid would be the responsibility of the Department of Defense rather than DOE; therefore, it was not considered.

The fifth level includes targets within a typical production plant or test area, such as:

- o Complete Weapons or Weapon Test Devices, Plutonium, or Highly Enriched Uranium in Weapons Components or Metal Parts,
- o Compounds Requiring Simple Processing to Fabricate Weapons Grade materials,
- o Compounds Requiring Complex Processing to Fabricate Weapons Grade Materials,
- o Dispersable Radioactive Material,
- o Unique or Critical Assets Necessary for Continuity of Operations.

Four types of adversary tactics were considered at the sixth level of the hierarchy:

- o External Assault,
- o Deceptive Action,
- o Insider Only,
- o Stealth.

"External Assault" should be clear, recalling from earlier remarks that it would not be on the scale of an invasion by a foreign power. We made an assumption that an assault would always be aided by at least one cooperating insider. "Deceptive Action" is defined as a tactic carried out by falsification of documents and other steps making the entry of an adversary into a site appear legitimate. "Insider Only" is self explanatory. "Stealth" is a tactic in which an attempt would be made to enter and leave a site surreptitiously.

At the seventh level, six countermeasure functions are identified:

- o Reduce Likelihood of Attempted Adversary Action by an Insider,

- o Delay Adversary's Access to, and Egress from, Targets,
- o Reduce Likelihood of Attempted Adversary Action by an Outsider,
- o Respond to Potential or Apparent Adversary Action,
- o Detection at Vital Internal Areas,
- o Detection at Perimeter.

The countermeasure functions are performed by security subsystems, considered at the eighth level of the hierarchy, such as:

- o Passive Deterrence/Defense,
- o Protective Force,
- o Physical Security,
- o Command, Control, and Communications,
- o Personnel Security,
- o Material Control and Accountability,
- o On-Site Transportation.

"Passive Defense" features are those designed and integrated into a process or weapons to make them inherently less attractive, accessible, or vulnerable. For example, a finished product might be immediately encased in a heavy shipping container that is difficult to remove and handle. Passive defense countermeasures are developed and carried out not by security personnel, but by product designers, plant engineers, and plant operations personnel.

"Passive deterrence" activities indirectly dissuade potential adversaries from taking action. Examples are public notices as to laws, penalties, and rewards. In addition, passive deterrence includes intelligence gathering and attempts to detect conditions that justify alerts, such as the theft of a large quantity of explosives in the immediate vicinity of a facility.

"Protective Force" refers to the on-site force that provides routine guard services, emergency response teams, and the management of both.

"Physical Security" consists of such components as fences, radiation monitors, and closed circuit television, all of which are intended to delay, detect, and, in some cases, assess the actions of a potential adversary.

"Command, Control, and Communications" is the subsystem intended to provide timely and effective management response and resource deployment in the event of an attack.

"Personnel Security" comprises such measures as background investigations, personnel security clearances, employee access controls, security awareness training, and the "two-person rule" (e.g., one person is never left alone with a nuclear warhead).

"Material Control & Accountability" encompasses a broad range of safeguard measures, each designed to either deter, defend against, or detect the theft or diversion of special nuclear materials. The material control aspect refers to control of the flow of the actual physical material in the course of a facility's operations. Material accountability is the system whereby nuclear material is measured and tracked by location as it progresses through a facility.

"On-Site Transportation" refers to systems for movement of SNM within the boundaries of a single facility.

Level 9 of the hierarchy is not discussed in further detail in order to avoid releasing confidential information. Given the above-mentioned hierarchy, security at each of the four sites was evaluated, one site at a time.

5. EXPERIMENTAL RESULTS

In this section, we present simulated results, which approximate the actual application, in order to avoid releasing classified information. The simulated results (see Figure 2) allow us to give an idea as to the actual importance of the elements at each level of the hierarchy, except for Levels 5 and 9 which are omitted. The reader should observe that, at each level, the weights sum to one. These weights were calculated automatically by the Expert Choice software package.

In addition to the hierarchy representing the security of a site, a related hierarchy, concerning the intersite transportation system, was also created and tested in Phase 1 using Expert Choice. For both hierarchies, sensitivity analysis was performed to observe how weights would change if the relative importance of some of the criteria were to change. A major benefit of this type of analysis would be in site-specific cases where facility managers could analyze the effect of changes in their assumptions regarding potential attack scenarios.

6. CONCLUSIONS

The original plan was for a Phase 2 study to follow Phase 1. As it turned out, the project was terminated at the completion of Phase 1. Despite this fact, the project was extremely successful. Based upon the AHP model, the assessment team made 92 major recommendations for security improvement. AHP was used to rank and weight each recommendation. DOE managers were highly responsive. In some cases they took action before the inspection team left the sites. Other corrective action required planning, budgeting, capital improvements, etc. Practically all of the recommendations were eventually implemented.

Thus, the Analytical Hierarchy Process was used to decompose the complex problem of security at nuclear facilities into its basic elements and help determine the requirements for countermeasures in reducing the likelihood of a successful attack.

Figure 2. Simulated Results

Level	Elements	Weights
2	Consequence Area	
	National Security	0.49
	Public Health and Safety	0.33
	Political and Social	0.18
3	Adversarial Strategies	
	Theft of Weapon	0.35
	Detonation of Weapon	0.30
	Theft of SNM	0.24
	Radiological Sabotage	0.11
4	Adversary Type	
	Terrorist (with insider assistance)	0.48
	Disgruntled Employee	0.17
	Criminal (insider or outsider)	0.12
	Foreign Agent	0.12
	Anti-Nuclear Extremist	0.11
6	Adversary Tactics	
	External Assault	0.38
	Deceptive Action	0.30
	Insider Only	0.28
	Stealth	0.04
7	Countermeasure Functions	
	Reduce Likelihood of Attempted Adversary Action by an Insider	0.24
	Delay Adversary's Access to, Egress from, Targets	0.22
	Reduce Likelihood of Attempted Adversary Action by an Outsider	0.21
	Respond to Potential or Apparent Adversary Action	0.19
	Detection at Vital Internal Areas	0.10
	Detection at Perimeter	0.04
8	Countermeasure Subsystems	
	Physical Security	0.23
	Passive Deterrence/Defense	0.20
	Protective Force	0.20
	Command, Control, and Communications	0.19
	Personnel Security	0.08
	On-Site Transportation	0.05
	Material Control and Accountability	0.05

The approach appears useful as a tool for planning future security requirements and resource allocation as well as for assessing the adequacy of security at existing facilities.

AHP IN PRACTICE: APPLICATIONS AND OBSERVATIONS FROM A MANAGEMENT CONSULTING PERSPECTIVE

Kenneth H. Mitchell
MicroAnalytics, Ltd.
1986 Queen Street East
Toronto, Canada M4L 1J2

Edward A. Wasil
Kogod College of Business Administration
American University
Washington, D.C. 20016

ABSTRACT

Over the last eight years, the Canadian management consulting firm of Woods Gordon has used the Analytic Hierarchy Process to help clients in the public and private sectors structure and solve complex, real-world decision problems. Many of the applications were costly projects that required detailed hierarchies with a large number of criteria and alternatives and that often involved a group of decision makers.

In this paper, we review the use of the AHP in a consulting environment and focus on the many practical considerations that users must address in order to facilitate a successful decision-making process. To provide some background on how the AHP works in practice, four applications are described in detail: a hospital's building and renovation program, strategic planning for information systems needs, contractor selection, and the allocation of military maintenance work.

1. INTRODUCTION

In the ten years since its introduction, the AHP has been used by decision makers to gain insight into a wide variety of complex, costly, and important decision problems. The chapter by Golden, Wasil, and Levy in this volume catalogs over 150 papers that apply the methodology in 29 different areas, ranging from health care to space exploration. While many of these papers apply the AHP to real-world problems, few provide detailed insights into the practical considerations that users (such as consultants) must address in order to facilitate a successful decision-making process. The goal of this paper is to review the decision-modeling process carried out by the management consulting firm of Woods Gordon when applying the AHP to help clients successfully solve problems in both the public and private sectors.

The next section in this paper presents some background on Woods Gordon and the steps they employed to model a decision problem. To provide some background on how the AHP works in practice, four real-world applications are described in detail: (1) a hospital's building and renovation program, (2) strategic planning for information systems needs, (3) contractor selection, and (4) the allocation of military maintenance work. Based on the experiences derived from using the AHP in a consulting environment, several observations are then made that can help promote successful applications of the process. The paper concludes with a discussion of some trends in the use of the AHP as a decision-modeling tool.

2. BACKGROUND

Woods Gordon is the largest management consulting firm in Canada. It was started in the early 1930s by two practitioners of industrial engineering techniques with the support of the Canadian industrialist who owned the factory in which they had developed these methods. Woods Gordon continued to have an active practice in quantitative methods as the firm expanded into other areas of marketing, economic, and financial consulting. In the 1960s, an operations research group was established that grew to a staff of about ten by the early 1980s.

In 1981, one of the senior consultants at Woods Gordon attended a seminar on the AHP and after a few weeks received a consulting request from one of the firm's regular clients that seemed appropriate for the AHP. A method was needed to quickly analyze a complex problem that involved assessing impacts, benefits, and prospects for nuclear power. After some discussion, the client and Woods Gordon agreed to share the cost of a trial application of the AHP. A group of knowledgeable experts was convened and the decision-making session was conducted by Professor Thomas Saaty, assisted by Woods Gordon consultants.

This first application showed that the AHP was a useful method for attacking complex decision problems. Also, the Woods Gordon staff gained valuable practical experience in applying the method. Soon thereafter, the firm submitted a competitive proposal that would use the AHP to develop a "scoring procedure" to select government contractors. The problem setting included multiple objectives and several decision makers. Woods Gordon was awarded the contract and the firm was thus launched into a new area of consulting practice.

From 1981 to 1985, Woods Gordon used the AHP to model problems for a wide variety of clients. Table 1 lists 15 different AHP applications that range from setting priorities for funding scientific research to assisting a town council in selecting a computer vendor. Many of the projects had a very large decision impact and nearly all were highly important to the client organization--often the most important project undertaken in several years. Each project involved a group of decision makers, with group sizes up to sixty. Often there were a large number of decision criteria that could be accommodated because of the hierarchical organization into levels and the grouping of subcriteria under major criteria. In those projects with a large number of alternatives, the lowest level of the hierarchy would contain scorable attributes on which the decision alternatives

Table 1. Applications of the AHP by Woods Gordon

Client	Year	Decision Impact*	Project Cost	AHP Modeling Cost	Group Size	Problem Size Criteria	Problem Size Alternatives	Problem Description
Government	1981	NA	$15,000	$15,000	6	45	4	evaluate nuclear power options
Hospital	1982	$25	60,000	20,000	8	9	42	set priorities for renovation project
Government	1982	100	50,000	30,000	10	75	UL	scoring system for selecting contractors
Financial Institution	1983	10	25,000	25,000	12	25	3	select EDP system and vendor
Military	1983	20	70,000	30,000	50	60	200	allocate equipment maintenance to contractors
School District	1983	NA	5,000	3,000	5	15	4	select senior administrator
Science Foundation	1983	NA	40,000	40,000	10	40	30	set priorities for funding scientific research
Government	1983	NA	18,000	18,000	30	40	NA	develop principles for regulating trucking industry

*In millions NA Not Available UL Unlimited

Table 1. (Continued)

Client	Year	Decision Impact*	Project Cost	AHP Modeling Cost	Group Size	Problem Size: Criteria	Problem Size: Alternatives	Problem Description
Teaching & Research Hospital	1983	NA	35,000	35,000	8	30	25	strategic planning for future program specialization
Government	1984	NA	10,000	10,000	3	75	UL	update criteria weights for contract selection
Hospital Group	1984	NA	40,000	15,000	8	15	18	computer strategy for a group of four hospitals
Energy Department	1984	NA	25,000	25,000	12	25	35	contingency plans for energy shortages
Disease Control Agency	1984	NA	20,000	20,000	15	12	20	choose alternatives for AIDS response
Postal Service	1984	50	75,000	75,000	60	40	100	set capital project priorities
Municipality	1985	1	8,000	8,000	12	15	2	assist town council in selecting computer vendor

*In millions NA Not Available UL Unlimited

could each be classified without having to explicitly compare one alternative to the others.

Woods Gordon selected the AHP to model the decision problem in each of these 15 applications. The AHP was viewed as providing a comprehensive, logical, and structural framework that enabled decision makers to focus on the problem of interest. In many of these applications, complex factors, qualitative considerations, and conflicting interests among the participants were involved. Essentially, the AHP helped decision makers to:

* identify the critical elements and issues of a problem,
* eliminate less important elements early in the decision-modeling process,
* solve problems in a timely manner, and
* inspire confidence and commitment among participants.

The hierarchical structure of the process seemed to correspond nicely with the way in which many decision makers personally sort out multi-factor problems. The process used only relevant, available information and did not need the "hard" data required by many structured decision methods. Many structured decision methods (e.g., mathematical programming models) require the factors in the decision--variables, criteria, and alternatives--to be scaled on a common scale (money being the most often used value scale) to which arithmetic operators can be applied. This usually results in using only production or financial data to model a problem. In many cases, this type of data may be difficult to collect or forecast. On the other hand, the AHP incorporates the comparative scaling of all decision factors as part of the hierarchical model. Any production or financial data that is a natural part of the decision problem can be expressed in broad levels or ranges and can be easily included in the model along with other qualitative factors.

The insight and clarity produced by structuring the problem in terms of its important factors coupled with the ease of making judgments about the importance of factors made the AHP easy to apply. Decision makers viewed the AHP as an efficient and cost-effective way to make a difficult decision or to devise a plan. It also allowed them to formally document their decision-making rationale. In summary, the AHP was looked upon as a straightforward decision-modeling approach that harnessed the judgments of experienced people in a structured way.

3. STEPS IN USING THE AHP

In using the AHP, a single decision maker or a group usually proceeds from problem identification, through an assessment of alternatives, to the final selection of a course of action. The overall process, which the consulting team at Woods Gordon used, consisted of the following ten steps.

Step 1. Define the Problem

It is important to establish whether the decision problem concerns allocating resources, choosing the best alternative, or planning a future course of action. Once this is clearly defined, the consulting team could provide advice about a workable structure for the problem. A workable structure must cover all factors and must properly relate alternatives to objectives, possibly through an intermediate level of criteria. Members of the consulting team with experience in applying the AHP and an MS/OR background usually possessed the skill required to determine the general outline of a hierarchy that was appropriate for a problem.

Step 2. Select the Decision Group

The consultants assisted the client in selecting a decision-making group. This group could include senior decision makers, technical staff, and advisors (from outside the department or organization) with an external perspective about the problem. It is important to select a group which collectively has the necessary expertise and information to "attack" the decision problem. Sometimes external advisors would be clients of the organization which the decision is intended to benefit. Furthermore, if the decision must stand up to outside scrutiny, it may be essential to include potential adversaries in the group. Typically, groups consisted of six to twelve members but larger groups were also used. We point out that the decision-making process was more "efficient" (mainly faster decisions) in the smaller groups (12 members or less). However, "effective" decision making (acceptance of the final decision and easy implementation) often required a large group so that all stakeholders could be represented.

Step 3. Identify Issues and Objectives

The first task of the group is to identify the issues and objectives which need to be considered in the decision. This step is important since it solidifies the group with a specific decision-making, problem-solving focus. The consultants assist by facilitating the identification of important factors, by recording them, and by preparing draft definitions of them in written form. The consultants are continually funneling written materials to the group so that details are documented and potential problems over semantic issues are averted. This allows the group to focus on the task of generating pairwise comparison matrices for the criteria and alternatives.

Step 4. Develop the Structure of the Hierarchy

The decision-making structure is an organized, hierarchical depiction of the decision problem in terms of the choices to be made, the objectives that are to be pursued, and the interests and criteria that must be taken into account. It is constructed by the group with the help of the consultants.

Step 5. Judge the Importance of the Decision Factors

Once a workable decision-making structure has been established (i.e., a structure that covers all factors and models the problem), the next step is to judge the importance of the decision factors. Factors are evaluated in a pairwise manner. Decision makers are asked to verbally judge two elements in terms of their importance to the objective to which they contribute. The verbal judgments are then converted to a numerical scale and entered into a computer program that performs the mathematical calculations and produces a set of weights. Woods Gordon developed a mainframe computer package which is used on time sharing for very large hierarchies. This program contains several features designed to tally scores for projects with a large number of alternatives. The Expert Choice microcomputer package is used for problems that contain a smaller number of criteria and alternatives. We should point out that factors shown to have little importance can be dropped at this stage of the process.

Step 6. Evaluate Alternatives

The decision alternatives that appear at the bottom level of the hierarchy are compared by the decision-making group.

Step 7. Report on Results

The judgments of the group are synthesized and the overall priorities of the alternatives are calculated.

Step 8. Check Reasonableness

The group must allow time for consideration of the reasonableness and implications of the AHP results. It is possible that some of the results may not seem appropriate once the decision makers have "stepped back" from the element by element evaluation and have had time to consider the results of the process. It is important to provide an opportunity for the decision makers to think through and understand the implications of the decision and, if necessary, to revise the process. For example, a revision is necessary when a key decision factor has been omitted or the hierarchical structure fails to take into account an important consideration. The group would then "backtrack," that is, return to earlier steps in the decision process and revise judgments, factors, or alternatives. This step also allows the group to recover from any "buyer's remorse" about the final decision. The consultants were always willing to address any criticisms about the analysis so that the group would realize that the final results were sound and not the result of some mysterious decision process.

Step 9. Finalize Choices

Once the reasonableness of the results has been carefully checked, the decision-making group convenes to resolve any outstanding issues and to finalize their decision.

Step 10. Documentation

The final step involves documenting the decision-making process. It helps to reinforce the soundness of the approach and also allows the group to easily review the process should the problem's characteristics change. A formal report would be drafted that states the definitions of decision factors, includes judgments about the factors and their importance, and documents the underlying rationale of the selection process. The report ranges in length from 20 pages to over 200 pages depending on the complexity of the project. Initial drafts of the report are circulated during the actual decision-making process and the report continues to evolve in the form of handouts for the workshop sessions. After the decision process is finished, the document usually requires only minor editing before it is completed.

A consulting project that followed these ten steps would last about eight weeks. A sample schedule that gives typical durations for each step is shown in Table 2. This schedule assumes three workshop sessions with the decision group. The first workshop is held in week 2 or 3 to identify broad issues and objectives. The second workshop, which occurs in week 4, is the main decision-making session in which the group would develop the hierarchical structure, judge the importance of the decision factors, and evaluate the alternatives. For a problem with a large number of criteria and alternatives, this session could span two days. The final workshop occurs in week 7 and is devoted to reviewing the results of the group's deliberations and to finalizing the group's decision.

4. APPLICATIONS OF THE AHP

In this section, we present four real-world consulting projects that were carried out by Woods Gordon and that used the AHP as the primary decision-modeling tool. Other consulting work was carried out in these projects and other management science methods were also used.

4.1 Facilities Planning

A large Canadian hospital needed to decide which facility improvements should be included in a building and renovation program. A preliminary study had identified sixty important departmental needs not all of which could be funded by the $25 million available for the project. Since all sixty were initially deemed equally important, the hospital had to employ a fair comparison method that would establish individual priorities. Using the AHP, a consulting team assisted the hospital's Steering Committee to develop these priorities. The Steering Committee was composed of seven medical and administrative managers, including the medical chiefs of staff, the director of nursing, the director of planning, and the executive director of the hospital. The consulting team included two management science consultants knowledgeable in the AHP and three facilities planning experts who

Table 2. Sample Consulting Schedule

Step	Description	Week
1	Define the Problem	1
2	Select the Decision Group	2
3	Identify Issues and Objectives	2-4
4	Develop the Structure of the Hierarchy	2-4
5	Judge the Importance of the Decision Factors	4
6	Evaluate Alternatives	4
7	Report on Results	5-6
8	Check Reasonableness	7
9	Finalize Choices	7
10	Documentation	8

surveyed the hospital's needs at the outset and developed final facility plans after the AHP decision-making sessions were completed. The entire project lasted about three months.

4.1.1 Developing a Decision Structure

In the first meeting with the committee, the consultants introduced the AHP approach. A number of important objectives were discussed and the initial, overall structure of the decision hierarchy was agreed upon. At this point, the consultants assisted the group in determining the hospital's service area needs (i.e., alternatives), in developing the relationships of needs to objectives, and in finalizing the decision hierarchy. This information was compiled by the consultants and incorporated into a written document that would form the basis for priority setting. The decision hierarchy is shown in Figure 1. This hierarchy has been slightly modified to protect the confidentiality of the client. From this figure, we see that the overall focus of the project was to determine the best use of the building and renovation funds, that is, the decisions are made in context, not in pursuit of the abstract goals of the hospital. The next level contains the relevant objectives. These were identified by asking questions such as "What are the real purposes or benefits that the major renovation project is supposed to achieve?" Two of the objectives were further decomposed into subobjectives. The alternatives for spending the funds are listed at the bottom of the hierarchy. Each of the alternatives was described in a paragraph that indicated the particular needs of a department or service of the hospital. For example, the nephrology department might request "a doubling of the department's space to relieve overcrowding." Cardiology might want to "move surgeons' offices closer to surgery units and improve the quality of the space." Several departments (e.g., respirology) had two specified needs. We point out that only some of the alternatives were determined to significantly relate to each objective. For example, only five service area needs relate to quality of patient care. This is illustrated by the letter codes shown alongside the line descending from the first

Figure 1. Hierarchy for Hospital Facilities Planning

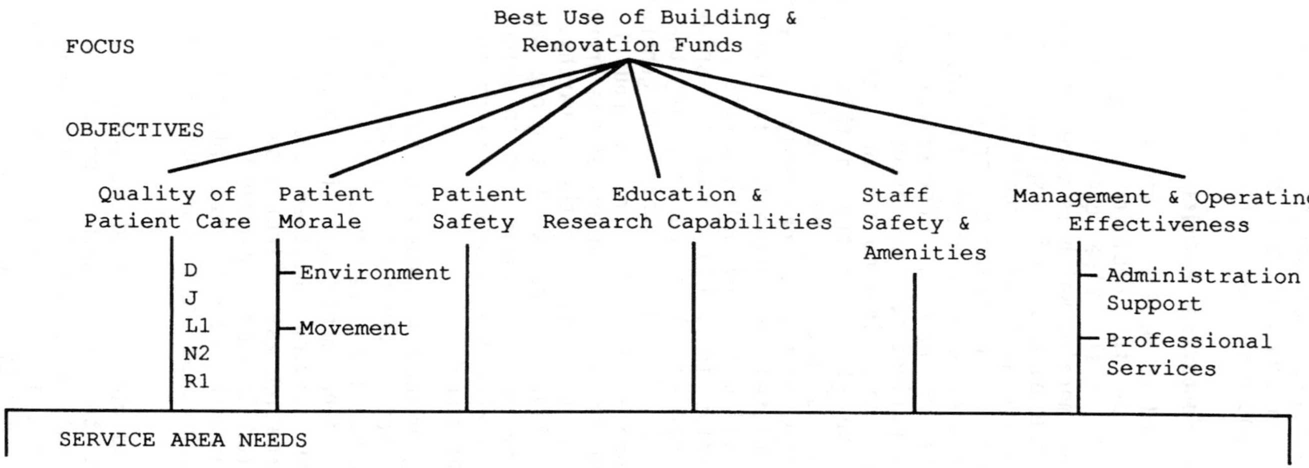

A	Anesthesia	L1	Respirology	U1	Rehab. Prosthetics	BB	Financial	NN	Surgery Offices
B	Allergy	L2	Respirology	U2	Rehab. Prosthetics	DD	Accounting	PP	Diag. Radiology
C	Cardiology	M2	Rheumatology	W	Neurosurgery	EE	Duty Rooms	QQ	Nursing Admin.
D	Cardio. Lab	N1	Obstet./Gyn.	AA1	Nursing Units	FF	Personnel	RR	Administration
E	E.C.G.	N2	Obstet./Gyn.	AA2	Empire Wing	HH	Psychology	SS	Shopping Mall
F3	Endo./Met.	O1	Ophthalmology	AA3	Watkins Wing	II	Purchasing	TT	Prefer. Accom.
I	Haem./Onc.	Q	Pathology	AA4	Angada Wing	KK	Admitting	UU	Day Care Center
J	Nephrology	R1	Neonatology	AA5	Victory Wing	MM	Emer. Med.	VV	Res. & Education
K1	Neurology	R2	Neonatology						

objective. The alternatives that specifically relate to the subobjectives (such as environment and movement) and to the remaining objectives are not displayed in Figure 1.

4.1.2 Developing Pairwise Comparison Matrices

The first author acted as moderator and was responsible for eliciting the judgments of the decision-making group. The participants responded to qualitative questions such as, "Of patient safety and patient morale, which is more important, and by how much, to achieving the best use of hospital funds?" Verbal responses on the qualitative scale of "equal" to "absolute" were then converted to the usual 1 to 9 numerical scale. To illustrate the process, the pairwise comparison matrix for the six second-level objectives is shown in Figure 2. For example, we see that improving the quality of patient care was judged to be absolutely more important than improving patient morale and therefore a value of 9 was assigned to the (I, II) entry in this matrix. The pairwise comparison data was entered by on-site terminal into the mainframe computer package mentioned previously. The resulting weights were then calculated and presented to the group for review. (This was an efficient and quick process. While one consultant recorded entries in a blank matrix displayed by an overhead projector, another consultant was entering the judgments into the computer terminal. Once the weights were generated, they were displayed by the projector to the entire group.) We point out that considerable debate amongst members of the group occurred when generating entries at this level of the hierarchy. However, by considering the objectives in pairs, the group was able to reach a consensus. There were no comparisons on which agreement could not be reached and when the results were examined, a high degree of consistency was found.

The decision-making group next generated judgments concerning the alternatives at the bottom level of the hierarchy, i.e., the service area needs. These needs had been identified through a review of the clinical departments and the administrative and support services. This review followed traditional facility planning methods. Information was collected on items such as the amount of space allocated to each department. This information gathering led to an interim report that identified the most important needs of each department for new space or space improvements.

The review described above provided an information database that allowed the group to judge whether or not a particular service need would contribute to the fulfillment of each objective. At this point, the service needs were compared with respect to the six objectives and the composite weights calculated. A partial list of these weights is shown in Figure 3. They were accepted by the group and recommended for adoption.

4.1.3 Implementation

After the weights were obtained, the project reverted to more traditional facilities consulting. The cost of new facilities and

Figure 2. Second-Level Pairwise Comparison Matrix for Hospital Facilities Planning

How much more important is Objective (row) than Objective (column) in providing the BEST USE OF FUNDS?	I	II	III	IV	V	VI	Weights
I. Quality of Patient Care	1	9	2	1	7	2	.286
II. Patient Morale	1/9	1	1/5	1/6	2	1/5	.039
III. Patient Safety	1/2	5	1	1/5	3	1/5	.101
IV. Education & Research Capabilities	1	6	5	1	7	2	.315
V. Staff Safety & Amenities	1/7	1/2	1/3	1/7	1	1/7	.032
VI. Management & Operating Effectiveness	1/2	5	5	1/2	7	1	.227

Consistency Ratio .09

the cost of renovating the old facilities were estimated for each service area need. Various building and renovation packages were designed using different combinations of new and renovated hospital space. Each of these packages included as many of the highest priority service area needs as possible within the $25 million spending limit. There was very little difference in the selection of which needs were served by the different packages. Rather, they represented different arrangements of the new physical facilities to serve the needs. One of the packages was selected for implementation.

We point out that although the group's priorities were accepted by the hospital's medical advisory committee, they were questioned by the board of directors. Since the group had documented the process, it was fairly easy to defend the results. The members of the group were committed to the final results--they felt that the process was fair and that their judgments and preferences were accurately modeled.

4.2 EDP Systems Development Strategy

This application illustrates how the AHP helped to model a difficult decision problem concerning electronic data processing (EDP) systems. Woods Gordon was engaged to help develop a long-range strategic plan to install management information systems in four jointly administered health care institutions. First, a team of computer consultants worked with a hospital task force to document the current computer situation and identify computer applications that would serve the needs of the four institutions.

Figure 3. A Partial List of Composite Weights for Hospital Facilities Planning

Service Area Needs	Quality of Patient Care	Patient Morale Environment	Patient Safety	Staff Safety & Amenities	Management & Operating Effectiveness Support	Management & Operating Effectiveness Services	Total
A Anesthesia						.008	.008
B Allergy						.008	.008
C Cardiology				.001		.003	.004
D Cardio. Lab.	.042						.042
E E.C.G.				.002	.004		.006
I Haem./Onc.		.002	.003				.005
J Nephrology	.104	.007	.026	.006	.001		.144
K1 Neurology						.008	.008
L1 Respirology	.018	.001	.007				.026
L2 Respirology						.008	.008
M2 Rheumatology						.008	.008
N1 Obstet./Gyn.						.009	.009
N2 Obstet./Gyn.	.042					.012	.054
O1 Ophthalmology						.004	.004
Q Pathology				.007			.007
R1 Neonatology	.080						.080
R2 Neonatology			.026				.026

Columns of weights for Movement subobjective and Education & Research objective are not displayed.

Then the AHP was used to develop priority information about the preferences and needs of the health care institutions for the information systems in various areas. The hierarchy for this application is shown in Figure 4. At the top level is the overall purpose or focus of the planning: To achieve the best computer strategy. The objectives are those purposes and benefits which the four hospitals desire as a result of adopting and implementing new management information systems. The criteria are the specific types of performance benefits that can be gained from a computer system in a hospital environment. Application areas, which can be chosen to acquire or develop new computer systems, are the decision alternatives.

The hierarchy was evaluated in the usual manner. First, the importance of the objectives was assessed by pairwise comparisons. Next, those criteria which contribute to each objective were evaluated with questions such as "Of the criteria contributing to Service Delivery, which is more important to the objective, Reduce Clerical Efforts or Improve Accuracy, and how important is it?" Finally, the potential computer applications which would address each criterion were evaluated as to their strength of contribution to that criterion. The synthesis of weights produced a ranked list of computer applications that was used by EDP specialists to develop a long-range plan for installing application systems at each of the four health care locations. The plan describes which applications should be adopted and also proposes a time schedule for implementation.

4.3 Contractor Selection

The granting of contracts by government agencies can be controversial decisions. In a recent Canadian government contracting process, the cabinet minister in charge wished to minimize the amount of controversy associated with the decision. To help achieve this, a committee was appointed with the responsibility of selecting contractors. The composition of the committee was chosen to be as representative as possible of the relevant public and private sector interests. It contained eight members: two from the legislative branch of the government, senior bureaucrats from four government departments, a labor representative, and a small business lobbyist.

The committee decided to select the contractors based on a specific set of selection criteria. This was based upon the view that if a comprehensive, objective system that is largely independent of the committee could be devised, then the possibility of controversy would be reduced. Woods Gordon was engaged to develop a weighted scoring system that would result in the selection of contractors.

The committee viewed this as a complex problem with many interest groups and political factors to take into account, but they were committed to devising a fair selection method. In particular, the two politicians that served as co-chairmen found it difficult to attach weights to problem objectives and criteria.

Figure 4. Hierarchy for EDP Systems Development Strategy

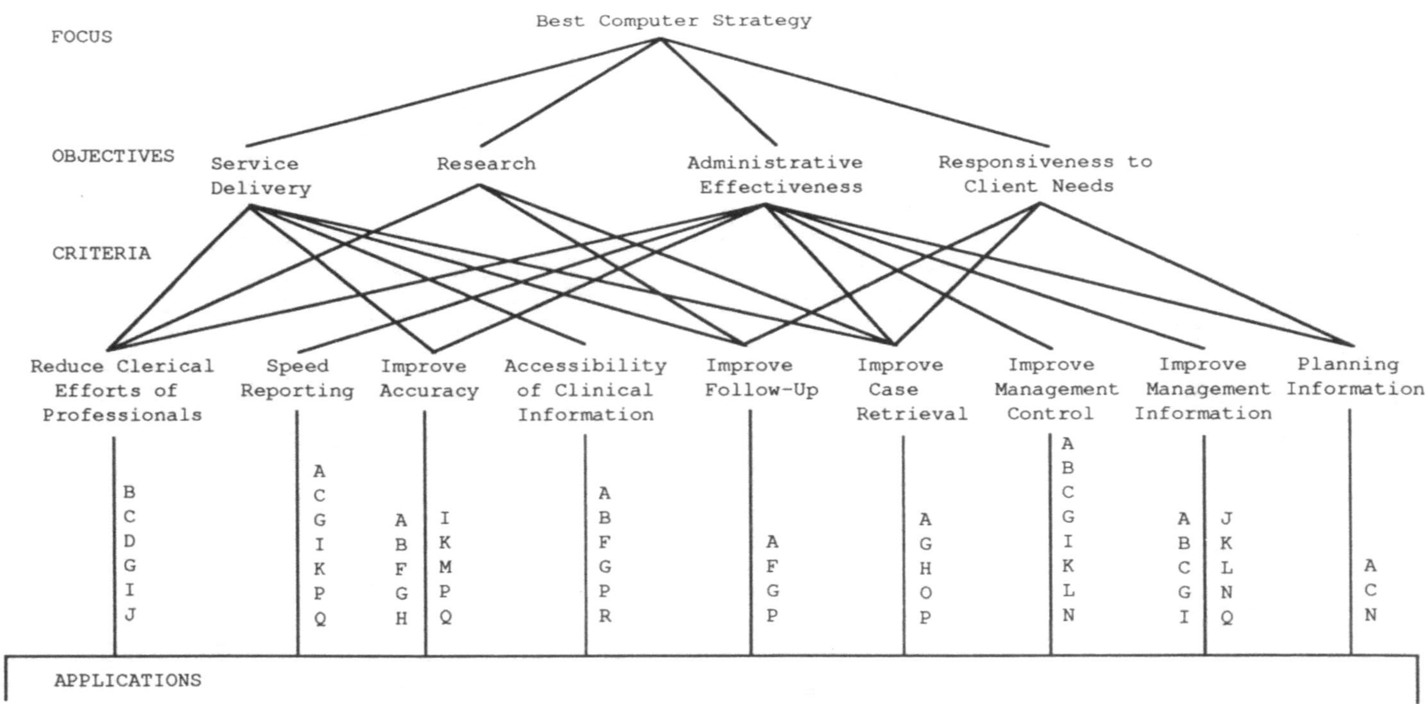

A Case Filing System
B Material Control System
C Service Support System
D Social Services System
E Client Information System
F Emergency Services System
G Production Laboratory System
H Indexing Storage Retrieval
I Charging Billing
J Budgeting System
K Purchasing and General Stores
L Work Orders
M Work Orders and Maintenance
N Management Reporting
O Research Support System
P Diagnostic Imaging
Q Patient Food Services
R Clinics System

When their opinions were solicited, speeches would ensue praising the importance of each factor--it seemed as if they felt that it was politically unwise to downgrade any factor.

The committee with the help of the consultants constructed the hierarchy shown in Figure 5. There were a total of seventeen selection criteria but only eleven are shown in this figure to protect confidentiality as this was a highly sensitive project. The criteria covered a wide range of concerns including the number of hours per day that the contractor was available for work, the experience of the contractor, and management capabilities. Each contractor was rated acceptable, good, or excellent and assigned points for each rating. The scoring system is shown in Figure 6. It was produced by judging the selection criteria in the usual manner and then assessing the importance of a contractor achieving an acceptable, good, or excellent score on a particular criterion. Notice that for the criterion amenities, excellent and good ratings receive the same score, whereas some criteria (e.g., finance and experience) have strong emphasis placed on achieving excellence. We point out that the scores shown in Figure 6 sum to 1000--this represents a weight of 1.0 in the hierarchy, multiplied by 1000 for convenience.

The scoring system was developed after 15-20 man-days of consulting and about 12 hours of the committee's time (six members of the committee participated). The AHP allowed the committee to resolve strong political conflicts among the members. The final results produced by the modeling process were accepted as a fair representation of the group's preferences.

At the implementation stage, each contractor completed a proposal form and was then graded on the selection criteria. (Applicants also had to meet minimum levels on certain criteria, such as finance, in order to be considered.) The proposal with the highest score was awarded the contract. This scoring system has been used for five years with one update of the criteria weights after two years.

4.4 Allocation of Military Maintenance Work

This application was part of a consulting project for the Canadian Department of National Defense (DND) that focused on the allocation of future land equipment overhaul among a government workshop facility and other Canadian industry sources. DND had just acquired a new tank and other new weapon systems were in the early stages of procurement. It was apparent that the maintenance needs of the new equipment could not be met without substantial new resources or reshuffling of existing resources. Certain equipment was best maintained in-house while some equipment could be fully or partially maintained by outside contractors. It was clear that the existing maintenance arrangements needed to be altered. Therefore, it was decided to re-plan the entire work allocation. The impact of this project is substantial: (1) several hundred man-years of Canadian employment are involved every year, (2) opportunities are created for Canadian industry to gain experience in new military technology, and (3) the military

Figure 5. Hierarchy for Selection of Contractors

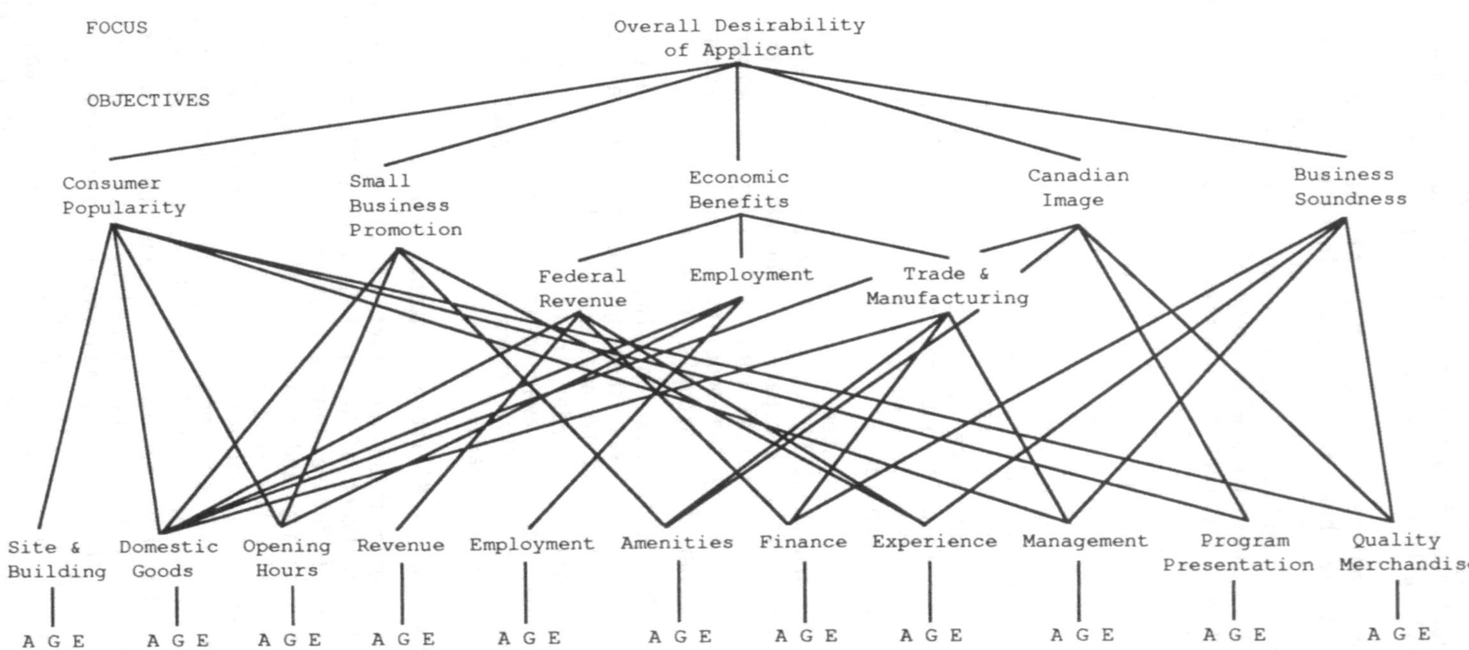

Figure 6. Scoring System for Selection of Contractors

Selection Criteria	Ratings		
	Acceptable	Good	Excellent
Site & Building	3	9	30
Domestic Goods	8	28	79
Opening Hours	28	51	92
Revenue	1	2	4
Employment	9	24	41
Amenities	3	7	7
Finance	10	36	61
Experience	11	34	81
Management	14	43	95
Program Presentation	6	11	20
Quality Merchandise	14	49	89

capability of the Canadian Armed Forces is influenced to a considerable degree. (The full details of this application can be found in the article by Mitchell and Bingham that appeared in the journal INFOR in 1986.)

Optimally allocating future maintenance work is a difficult process. Conflicts among military requirements, efficiency criteria, and industrial benefits objectives are difficult to resolve since the conflicts involve many qualitative issues. For example, the analysis might need to take into account qualitative factors such as the flexibility of military response, the best use of existing facilities, and the notion of fostering Canadian industry. The interests of DND, other government departments, Canadian industry, and the workforce at the government repair depot must also be considered.

A project team of DND staff and Woods Gordon consultants analyzed the overhaul workload and developed a detailed information package that outlined the required tasks and the appropriate workshops for maintaining 200 components and systems. This information was combined with the knowledge and experience of fifty senior military officers, industrial development specialists, and technical personnel to produce an objective-based scoring system. This system measures how well the objectives are satisfied if a contract is awarded to a specific contractor. The hierarchy for this application is shown in Figure 7. From this

Figure 7. Hierarchy for Measuring the Effectiveness of Work Allocation Decisions

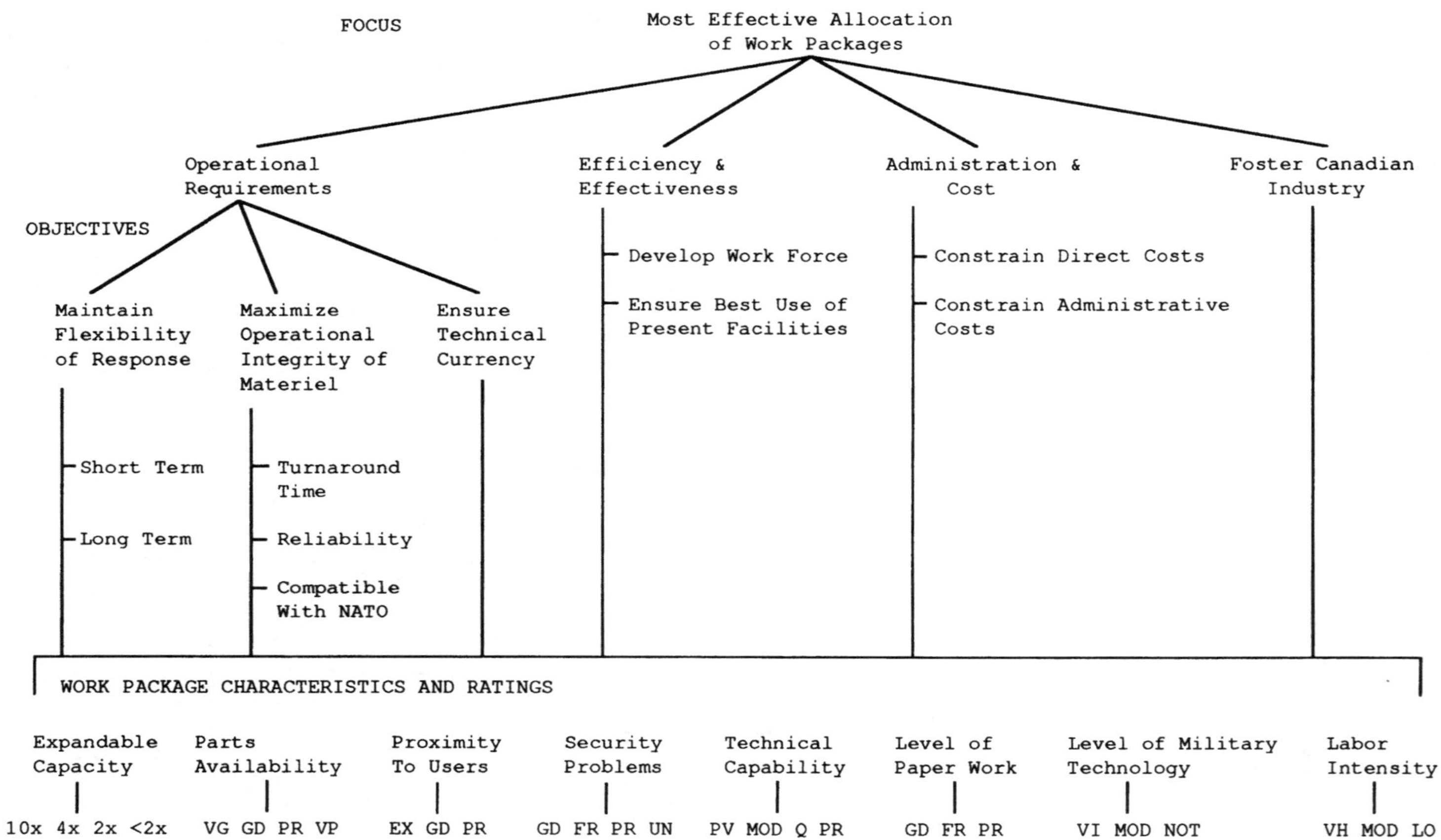

figure we see that the relevant selection criteria included parts availability, the proximity of the contractor to field users, security problems, and the contractor's technical capability.

In a second phase, these scores are used to determine the most effective allocation of all work packages. The government workshop is loaded to capacity and a straightforward mathematical programming model is solved to allocate the remaining work to alternate work sites.

In judging the importance of the objectives, the group agreed to recognize interests that were outside the DND. The senior ranking officer commented that, although the primary mandate of the Canadian forces is military, the priority setting also had to reflect the fiscal and operational constraints (requirements to contract out some work and follow "business logic" in running the operation) within which the forces are expected to operate in peace time.

Working through the decision-modeling process with a group of fifty participants (30 middle-level managers and 20 senior executives) was challenging, especially when eliciting judgments. The consultants were prepared for a large number of objections when the results were critically reviewed. To their surprise, most objections were matters of definition or clarification and they were resolved easily.

Using the AHP-based scores, the consultants were able to calculate a measure of overall program effectiveness for the current allocation of work and the new, recommended allocation. They found a 15% improvement in overall effectiveness against the multiple decision objectives. On a base of direct program spending of about $24 million per year, 15% would be equivalent to $3.6 million per year of improved value achieved from the recommended allocation of work.

5. AHP IN PRACTICE

In this final section, we focus on the practical considerations that users must address in order to facilitate a successful decision-making application with the AHP. These observations are derived from the many Woods Gordon consulting projects that used the AHP as the primary decision-modeling tool.

A key way of ensuring success is to establish certain project roles. The workshop leader is usually the AHP expert and he should be very comfortable with the process. The leader acts as a facilitator, that is, assisting the decision-making group in modeling the problem in the specified time frame. A consultant that is knowledgeable in a functional area of the problem (such as marketing or corporate planning) may also be involved in the process. These functional consultants provide help in defining and structuring the problem and documenting key points of the discussion. They usually require several hours of training to familiarize themselves with the AHP. A second AHP analyst is helpful in orchestrating the workshop and operating the on-line computer system (as described in the facilities planning application). Finally, a client coordinator can work with the consulting team to provide information about important issues and about the personalities of the decision group's members. This allows the consultants to be reasonably sure in advance that the process is structured in a manner that is appropriate to the group.

One of the most critical aspects in delivering the AHP to decision makers is the group process. A solution to a decision problem is essentially derived from the collective abilities of the group's members. The key to success is ensuring that the group accepts responsibility for reaching a final decision. Even if a problem is intractable so that no decision can be reached, the consultants can help the group gain insight into a problem by considering all of the important objectives, criteria, and alternatives. In this regard, it is important that the consulting team possess leadership as well as analytic skills.

The decision hierarchy should be appropriate for the decision-making group. It should be "theoretically correct" as well as workable for the group, that is, not so complex that judgments are difficult to make. It is also important for the group to formally carry out the AHP: Weights should not be assigned to criteria in an ad hoc way using a hierarchy that is not acceptable to the group. Successfully structuring the problem requires an organized project team that has ample preparation time (about 2 to 3 weeks of preparation was required in each of the projects described earlier). This will ensure that a balance is maintained between letting the group solve its own problem and following a logical decision process.

The AHP is an appealing tool for modeling a wide-range of decision problems. Before selecting the AHP as the primary modeling tool, users should keep in mind three limitations of the process. First, the AHP primarily aids the decision-making skills and thought processes of the group. In some applications, "clear thinking" is insufficient to make a decision; additional research and data gathering are also required. The AHP should not be used in such situations to replace the basic scientific method of constructing and testing hypotheses about system behavior. Second, it is not always advantageous to openly outline objectives, criteria, and alternatives in a group setting. A decision made by a single individual (perhaps arising from an application of the AHP) may be the only way to obtain action in some situations. Third, it is tempting to make a premature decision since the AHP allows a problem to be solved with whatever amount of information is available. Decision-making groups must be cautious and patient when gathering information that is pertinent to the problem. They must be sure that sufficient basic information has been obtained so that informed judgments can be rendered.

CHOOSING INITIAL ANTIBIOTIC THERAPY FOR ACUTE PYELONEPHRITIS

James G. Dolan
General Medicine Unit
Department of Medicine
Rochester General Hospital
Rochester, New York 14621

ABSTRACT

An AHP model was used to develop antibiotic treatment guidelines for young women hospitalized for acute pyelonephritis. Seven antibiotic regimens representative of current treatment recommendations were compared relative to four major criteria. The resulting analysis identified a combined regimen of ampicillin and gentamicin as the best choice for initial treatment pending results of urine culture and antibiotic sensitivity testing. The use of this regimen was recommended to a group of physicians. Subsequently the use of ampicillin and gentamicin increased significantly in young women with pyelonephritis. This study shows that a significant change in the process of patient care was associated with treatment recommendations based on the AHP. This finding indicates that the AHP may be a valuable tool for helping physicians make better, more logically consistent patient management decisions.

1. INTRODUCTION

Acute bacterial pyelonephritis is the term used to describe bacterial infection of the kidney. It is a potentially life-threatening illness because the infection can spread into the bloodstream and cause shock and other systemic complications. To reduce the chances of both systemic and local complications, prompt treatment is indicated with an antibiotic effective against the infecting organism.

The choice of antibiotic treatment for acute pyelonephritis is made under conditions of uncertainty because urine culture and sensitivity results are not available for approximately 48 hours. Thus, initial therapy must be based on an educated guess regarding the nature of the infecting organism and its antibiotic sensitivities. (Once the organism is identified and antibiotic sensitivities known, the initial treatment regimen can be adjusted accordingly.) Although most urinary tract infections are caused by a single organism, *E.coli*, other bacteria are found, especially in men and older women probably due to a higher prevalence of genitourinary abnormalities and co-existing illnesses. Moreover, knowing the infecting organism does not necessarily solve the problem of choosing the most appropriate antibiotic treatment since different isolates of the same bacterial species can differ in their antibiotic susceptibilities. (This is the reason most organisms isolated from an infected site are routinely tested against a variety of antibiotics.)

The situation is further complicated by the large number of antibiotics available. Currently, there are at least 30 different antibiotics, not including combination treatment regimens, that could reasonably be used for initial treatment of acute bacterial pyelonephritis. Although the differences in treatment efficacy among most of these agents is small, they vary considerably in other respects such as frequency of side effects and cost.

Faced with a complicated patient management situation, most clinicians rely on published expert recommendations to help decide the best course of action. In treating acute pyelonephritis, however, published recommendations are not that helpful. Recommendations vary from textbook to textbook and most are stated in general terms, recommending a class of antibiotics - which can contain up to 4 or 5 different drugs - instead of a specific treatment regimen that clinicians can order for their patients.

Antibiotic therapy has long been recognized as a particularly difficult area of clinical medicine and many attempts have been made to encourage physicians to follow sound antibiotic prescribing practices. For example, the Medical Letter recommends that antibiotics be chosen based on activity versus the infecting organism, expected toxicity, drug penetration into the infected site, patient specific factors such as allergies and co-existing illnesses, and cost [1]. However, few if any, of these previous attempts have been based on a formal analysis of the clinical decision facing the clinician. As part of a larger project to encourage cost-effective clinical practices at three hospitals in Rochester, New York, I used the Analytic Hierarchy Process to identify the best of seven currently recommended antibiotic regimens for initial treatment of acute pyelonephritis, made treatment recommendations based on the results, and assessed the recommendations' impact on clinical practice.

2. THE AHP MODEL

The hierarchy used for the analysis is shown in Figure 1. The goal was to determine the best initial antibiotic regimen for a young woman hospitalized for treatment of acute, uncomplicated pyelonephritis. Young women with pyelonephritis constitute a homogeneous group of patients with few additional complicating factors. The analysis was directed specifically towards these patients in order to simplify the clinical problem as much as possible. The alternatives consisted of the seven treatment regimens listed in Table 1. These alternatives include one member of every antibiotic class that was recommended at the time the model was built [5,6,7,8]. (Since the analysis was performed, three new antibiotics from different classes have become available.) It is assumed that the patient has no contraindications to any of these antibiotics. The decision criteria included in the model were based on accepted clinical management guidelines regarding selection of antibiotic therapy

Figure 1. Hierarchy for Best Initial Antibiotic Regimen (TMP/SMX = Trimethoprim/Sulfamethoxazole)

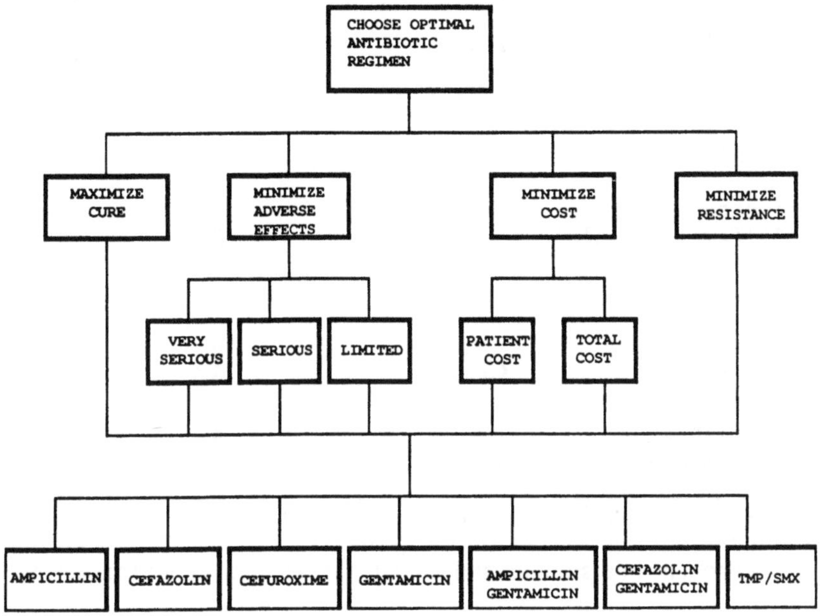

which are reflected in the definition and underlying rationale for each criterion listed below [1].

A. Maximize Cure. The antibiotic regimen chosen should be the one with the highest chance of curing the infection. In keeping with clinical practice, it is assumed that antibiotic sensitivities determined by *in-vitro* testing predict clinical cure of the infection.

B. Minimize Adverse Effects. The regimen chosen should be the one with the lowest expected rate of toxicity. Adverse effects were further divided into very serious, serious, and limited. Very serious adverse effects were defined as those which: (1) are life-threatening or likely to result in permanent deleterious effects on health or functional capacity if untreated, or (2) result in extremely distressing symptoms. Anaphylactic shock, hemolytic anemia, and decreased hearing are some of the adverse effects that were classified as very serious. Serious adverse effects referred to those which: (1) are potentially life-threatening or capable of having a permanent deleterious effect on health or functional capacity if untreated, or (2) result in distressing symptoms. Serious adverse effects included, among others, diarrhea, skin rash, jaundice, and vomiting. Limited adverse effects were defined as adverse effects which (1) are not life-threatening and (2) result in minor, easily-controlled symptoms. Nausea, asymptomatic abnormalities in biochemical liver function tests, and facial flushing are some examples of limited adverse effects.

Table 1. The Alternative Antibiotic Treatment Recommendations

Treatment Regimen	Dose	Frequency (hours)
Ampicillin	1 gram	Every 6
Cefazolin[a]	1 gram	8
Gentamicin[b]	80 mg.	8
Trimethoprim/ Sulfamethoxazole	160 mg.	6
Cefuroxime[c]	750 mg.	8
Ampicillin & Gentamicin	1 gram 80 mg.	6 8
Cefazolin & Gentamicin	1 gram 80 mg.	8 8

a--representing first generation cephalosporins
b--representing the aminoglycosides
c--representing second and third generation cephalosporins

C. **Minimize Cost.** The regimen chosen should be the one that costs the least. Treatment costs were further divided into patient cost, the out-of-pocket expense to the patient, and total cost, the actual cost of the antibiotic regimen to the health care system. For this analysis, relative antibiotic charges and costs were assumed to be equivalent.

D. **Minimize Resistance.** Strains of bacteria resistant to commonly used antibiotics can develop in different settings. The existence of resistant strains of bacteria in hospital environments makes the treatment of infections acquired in the hospital difficult. Thus, minimizing the development of bacterial resistance is a consideration whenever antibiotics are used in a hospitalized patient.

3. COMPARISONS

Weights for the decision criteria on the middle two levels of the hierarchy were based on pairwise comparisons made by 61 practicing clinicians from the Department of Medicine at Rochester General Hospital. The individual responses were combined by calculating the geometric mean for each pairwise comparison [2]; the mean values were then used to calculate the weights for each of the criteria. The results are shown in Table 2.

Table 2. Geometric Mean Comparison Judgments Made by 61 Clinicians and Resulting Criteria Weights

A. COMPARISON #1. BASED ON 57 RESPONSES.*

	MAE	Cost	Cure	Resist	Weight
Minimize Adverse Effects (MAE)	1	5.1	1/2.2	2.2	0.294
Minimize Cost (Cost)	1/5.1	1	1/5	1/2.1	0.073
Maximize Cure (Cure)	2.2	5	1	3.7	0.496
Minimize Resistance (Resist)	1/2.2	2.1	1/3.7	1	0.137

Consistency Ratio = 0.022

B. COMPARISON #2. BASED ON 59 RESPONSES.*

	VSSE	SSE	LSE	Weight
Minimize Very Serious Side Effects (VSSE)	1	4.6	6.8	0.698
Minimize Serious Side Effects (SSE)	1/4.6	1	5.6	0.236
Minimize Limited Side Effects (LSE)	1/6.8	1/5.6	1	0.066

Consistency Ratio = 0.173

C. COMPARISON #3. BASED ON 56 RESPONSES.*

	Tcost	Pcost	Weight
Minimize Total Cost (Tcost)	1	1.1	0.524
Minimize Patient Cost (Pcost)	1/1.1	1	0.476

Consistency Ratio = 0.0

* - Some of the physicians' comparisons were incomplete. The number of responses indicates the number of returned questionnaires that had a complete set of judgments for that comparison.

Maximize Cure was considered to be the most important criterion, followed by Minimize Adverse Effects, Minimize Resistance, and Minimize Cost. The clinicians made very consistent judgments among these major criteria as indicated by the excellent consistency ratio of 0.022. In contrast, the aggregated comparisons regarding minimizing the three types of adverse effects were much less consistent with a consistency ratio of 0.173. This relatively high degree of inconsistency was due primarily to the tendency of clinicians to rank minimizing both very serious and serious adverse effects very strongly to absolutely more important than minimizing limited adverse effects. It is unlikely that these inconsistencies adversely affected the analysis, however, because the resulting rank order and comparison weights assigned to minimizing the three types of adverse effects are quite reasonable and, as noted in the final results, the overall consistency of the analysis is very good.

Minimizing Cost was considered to be the least important of the major criteria and the clinicians considered minimizing total cost and patient out-of-pocket cost to be almost equally important. This is in keeping with the general notion that alternative treatments, especially those given for a short period of time for acute problems, should be compared primarily on clinical rather than economic grounds.

The weights for the seven antibiotic regimens relative to the criterion Maximize Cure were based on published data regarding the expected likelihoods of potential infecting organisms [4] and the results of routine antibiotic sensitivity testing at the Rochester General Hospital clinical microbiology laboratory. The probability of cure for each antibiotic regimen was calculated by multiplying the probability that each organism was susceptible times its expected frequency and summing the results for all potential bacteria. Comparisons among the seven regimens were made directly by normalizing the vector composed of the regimens' odds of cure. Odds of cure were used instead of probabilities of cure because of the possibility that the compression of likelihoods that occurs at the two extremes of the probability scale, 0 and 1, would distort the pairwise comparisons between regimens. These data are shown in Table 3.

The specific adverse effects associated with each antibiotic regimen and their expected frequencies were identified through a literature search. The adverse effects were then categorized as very serious, serious, or limited according to the above definitions. The likelihood of each type of adverse effect for each antibiotic regimen was then estimated by adding the lowest reported rates for the adverse effects in each category associated with the regimen. (The lowest rate was used because the expected duration of treatment was short, 48 hours.) These probabilities were converted to odds and the regimens compared using the same technique as that used for the Maximize Cure comparisons. Regimens with no reported limited adverse effects were arbitrarily assigned odds of 100:1 for comparative purposes. The results are shown in Table 4.

Table 3. Maximize Cure Data

Treatment Regimen	Odds of Cure*	Normalized Odds of Cure
Ampicillin	2.7	0.02
Cefazolin	5.0	0.04
Cefuroxime	24.1	0.18
Gentamicin	25.0	0.18
Trimethoprim/ Sulfamethoxazole	10.9	0.08
Ampicillin & Gentamicin	44.2	0.32
Cefazolin & Gentamicin	24.3	0.18

* - Odds expressed relative to 1, e.g., 2.7 = 2.7:1.

Table 4. Adverse Effects Data

	Odds (Normalized Odds) of Avoiding Adverse Effect[a]						
Adverse Effect	Amp	Cef	Cefu	Gent	Tmp/Smx	Amp & Gent	Cef & Gent
Very Serious	325 (0.26)	336 (0.27)	90 (0.07)	49 (0.04)	369 (0.29)	42 (0.03)	42 (0.03)
Serious	6 (0.09)	7 (0.11)	10 (0.15)	8 (0.12)	28 (0.43)	3 (0.05)	3 (0.05)
Limited	b (0.22)	21 (0.05)	12 (0.03)	b (0.22)	b (0.22)	b (0.22)	21 (0.05)

a - Odds expressed relative to 1, e.g., 325 = 325:1.

b - No reported Limited Adverse Effects. Odds of 100:1 were used arbitrarily.

Table 5. Antibiotic Total Costs (Cost of 2 Days Treatment at Rochester General Hospital in 1986)

Treatment Regimen	Charges	Reciprocal Charges	Normalized Reciprocal Charges
Ampicillin	$12.00	0.083	0.42
Cefazolin	$23.70	0.04	0.20
Cefuroxime	$33.96	0.03	0.15
TMP/SMX	$67.84	0.015	0.08
Gentamicin	$91.42*	0.011	0.06
Ampicillin & Gentamicin	$103.42*	0.0097	0.05
Cefazolin & Gentamicin	$115.12*	0.0087	0.04

* - Charge includes two serum drug levels at $41 each.

The charges for the antibiotic regimens at Rochester General Hospital during 1986, including both drug and administration fees, were used to compare the regimens in terms of Minimizing Total Cost. Regimens were compared directly by normalizing the reciprocals of their respective charges, since the goal was to minimize costs. The data are shown in Table 5. It was assumed that the patient had hospitalization insurance that would cover all antibiotic charges. Therefore, there was no difference among the regimens in terms of Patient Cost.

Comparisons among the treatment regimens in terms of Minimize Resistance were based on a consensus judgment made by three specialists in Infectious Diseases that all regimens were equal except cefuroxime, which they thought was moderately more likely than the others to induce resistance. The resulting comparison matrix is shown in Table 6.

Table 6. Minimize Resistance Comparison Matrix

	AMP	CEF	CEFU	GENT	TMP/SMX	A&G	C&G	Weight
Ampicillin (AMP)	1	1	3	1	1	1	1	0.158
Cefazolin (CEF)	1	1	3	1	1	1	1	0.158
Cefuroxime (CEFU)	1/3	1/3	1	1/3	1/3	1/3	1/3	0.053
Gentamicin (GENT)	1	1	3	1	1	1	1	0.158
Trimethoprim/ Sulfamethoxazole (TMP/SMX)	1	1	3	1	1	1	1	0.158
Ampicillin & Gentamicin (A&G)	1	1	3	1	1	1	1	0.158
Cefazolin & Gentamicin (C&G)	1	1	3	1	1	1	1	0.158

Consistency Ratio = 0.0

4. SENSITIVITY ANALYSIS

To determine the effect of changing some of the assumptions in the baseline model, the analysis was repeated to see how the results would change if:

Case 1. All antibiotic regimens were considered equal in terms of minimizing all three types of adverse effects.

Case 2. Routine serum gentamicin levels are not measured, thereby reducing the cost of the three regimens including gentamicin by $82.

Case 3. All regimens were considered equal in terms of minimizing resistance.

Case 4. Maximize Cure and Minimize Adverse Effects were considered equally important major criteria.

5. RESULTS

The analysis was carried out using Expert Choice [3]. The results are shown in Table 7. The baseline analysis identified ampicillin combined with gentamicin as the best treatment regimen with a priority score of 0.204 and an overall consistency ratio of 0.02. Ampicillin and gentamicin remains the most preferred therapy if all regimens are considered equal in terms of avoiding side effects, if gentamicin levels are not drawn, and if all regimens are considered equally likely to minimize the development of antibiotic resistance (Cases 1 - 3). When Minimize Adverse Effects and Maximize Cure are considered equally important criteria (Case 4), trimethoprim/sulfamethoxazole becomes the most preferred therapy with a score of 0.188 compared to 0.176 for ampicillin and gentamicin.

Table 7. Results

		OVERALL PRIORITIES			
		Sensitivity Analysis			
Treatment Regimen	Baseline	Case 1	Case 2	Case 3	Case 4
Ampicillin & Gentamicin	0.203	0.219	0.207	0.201	0.176
TMP/SMX	0.164	0.162	0.163	0.162	0.188
Gentamicin	0.141	0.157	0.150	0.139	0.129
Cefuroxime	0.132	0.131	0.129	0.144	0.123
Cefazolin & Gentamicin	0.128	0.129	0.129	0.126	0.114
Ampicillin	0.116	0.109	0.109	0.114	0.136
Cefazolin	0.116	0.094	0.113	0.114	0.134

6. IMPLEMENTATION OF MODEL RESULTS

The AHP model and its results were presented to physicians practicing at Rochester General at a weekly continuing education conference. Based on the AHP analysis, a combined regimen of ampicillin and gentamicin was recommended for initial treatment of young women admitted for treatment of suspected pyelonephritis.

The impact of the treatment recommendation was assessed by comparing the initial antibiotic regimens prescribed for patients hospitalized for treatment of urinary tract infections during the 25 months preceding the conference versus those prescribed during the 11 months after the conference. Patients were divided into three groups: young women, less than 50 years old, older women, and men. The latter two groups served as controls, since the treatment recommendations were made specifically for young women.

The results are summarized in Table 8. Following the presentation of the AHP model, the use of ampicillin and gentamicin increased significantly in the targeted patient group, young women, but did not change in either of the two control groups.

Table 8. Antibiotic Regimens Prescribed Before and After AHP-based Treatment Recommendations

Patient Group	Proportion Prescribed Ampicillin & Gentamicin Before Recommendations	After Recommendations	
Young Women	23/48 (48%)	19/24 (79%)	*
Older Women	28/89 (31%)	8/34 (24%)	**
Men	16/40 (40%)	9/18 (50%)	**

* - $z = 2.51$, $p < 0.01$ (one-tailed)

** - $p > 0.10$

7. CONCLUSIONS

The choice of antibiotic treatment is typical of many patient management decisions faced by clinicians every day. Numerous alternatives are available and the choice among them depends on multiple factors, some of which can be highly subjective. The AHP was expressly designed to help guide decision making in this type of situation and it proved to be an effective approach to this problem. The significant change in antibiotic prescribing behavior in the targeted patient group following the AHP-based recommendation, while not proving a cause and effect relationship, certainly implies that patient management guidelines based on an AHP analysis can influence clinical practice. This experience with the AHP suggests that it will prove to be a valuable patient management tool by enabling physicians to make better, more logically consistent patient management decisions.

8. REFERENCES

1. M. Abramowicz, editor, **Medical Letter Handbook of Antimicrobial Therapy**, The Medical Letter, Inc., New Rochelle, New York (1986).

2. J. Aczel and T. Saaty, "Procedures for Synthesizing Ratio Judgements," **Journal of Mathematical Psychology, 27**, 93-102 (1983).

3. E. Forman, **Expert Choice**, Decision Support Software, Inc., McLean, Virginia (1986).

4. R. Rubin, "Infections of the Urinary Tract," in **Scientific American Medicine**, E. Rubenstein and D. Federman, editors, Scientific American, New York, New York (1987).

5. J. Sanford, **Guide to Antimicrobial Therapy**, Antimicrobial Therapy, Inc., West Bethesda, Maryland (1987).

6. J. Sobel and D. Kaye, "Urinary Tract Infection," in **Principles and Practice of Infectious Diseases**, G. Mandel, R. Douglas, and J. Bennett, editors, John Wiley & Sons, New York, New York (1985).

7. W. Stamm and M. Turck, "Urinary Tract Infection, Pyelonephritis and Related Conditions," in **Harrison's Principles of Internal Medicine**, E. Braunwald, K. Isselbacher, R. Petersdorf et al., editors, McGraw-Hill, New York, New York (1987).

8. W. Valenti and R. Reese, "Genitourinary Tract Infections," in **A Practical Approach to Infectious Diseases**, R. Reese and R. Douglas, editors, Little, Brown & Co., Boston, Massachusetts (1986).

AN ANALYSIS OF CONFLICT IN NORTHERN IRELAND

Joyce M. Alexander
Department of Mathematics, Physics, and Computer Science
Immaculata College
Immaculata, Pennsylvania 19345

ABSTRACT

In this study, we model the Northern Ireland conflict using the Analytic Hierarchy Process. Our model is an update of earlier analyses carried out in 1976, 1977, and 1982 in which it was shown that the outcome which would most satisfy the aspirations of all parties would be legislative independence for Northern Ireland. The current analysis takes into account important changes that have taken place since the earlier work was performed. We show that the most satisfactory outcome is still one of legislative independence.

1. INTRODUCTION

One of the most interesting and provocative applications of the Analytic Hierarchy Process over the last twelve years has been in the area of conflict analysis and resolution. Many authors have used the AHP to model and propose solutions to major conflicts such as those in the Middle East, South Africa, the Falkland Islands, and Northern Ireland.

The methodology of the AHP is ideally suited for conflict analysis. In general, the conflict hierarchy consists of four levels--problem, actors, objectives, and outcomes. A general hierarchy is shown in Figure 1. It is necessary for decision makers to identify the parties to the conflict, the basic objectives that each party aims to achieve, and the basic political structures.

Several important conflict analysis studies are listed in Table 1. Some of these studies look ahead and seek to determine a preferred outcome. Others are retrospective in that they look back in order to better understand why certain outcomes emerged. In general, these are thought-provoking papers that represent rather creative AHP applications.

A collection of new case studies is forthcoming [10]. One of these new studies suggests possible negotiating strategies in the South African conflict. Another study is concerned with the recent free-trade negotiations between the United States and Canada.

As noted in Table 1, some of the earliest case studies using the AHP examined the conflict in Northern Ireland [1,3]. This analysis was later updated [2] to reflect a number of changes that had occurred since the earlier work. These studies showed that the political outcome which would best satisfy the aspirations of all parties to the conflict would be legislative independence for

Figure 1. General Conflict Hierarchy

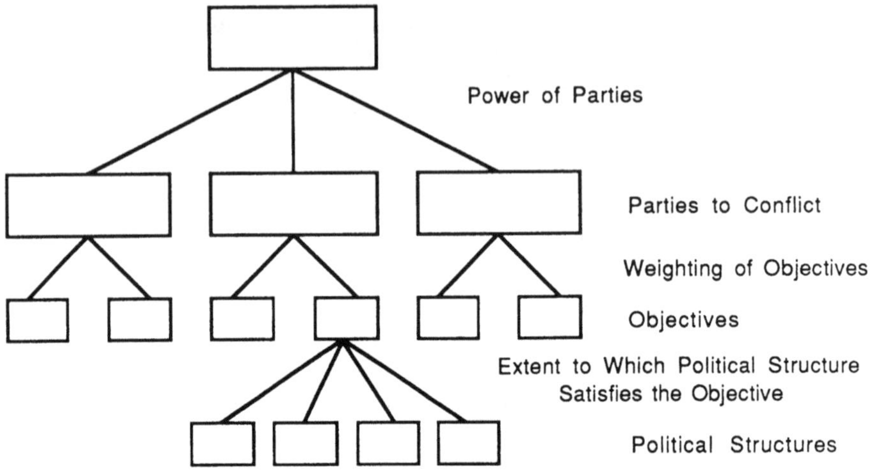

Northern Ireland. There was also a short-term compromise solution: Northern Ireland could be governed by an assembly, subordinate to the British government, but with a wide range of powers and a large measure of autonomy.

Since the 1982 analysis, there have been many significant changes. The present study was therefore carried out in July, 1988, to take these changes into account.

For this analysis of the Northern Ireland conflict, as in the earlier studies, the conceptual hierarchy follows very simple lines. However, this simple structure is enriched by an analysis of stability, including the use of the backward process to test the stability of the final outcome. There is also a study of how certain thresholds of power may affect the final result.

An important feature, of both this work and the earlier analyses, is the ongoing contact with participants in the conflict. The author visited Northern Ireland many times between 1975 and 1988 and obtained first-hand information and views from political and community leaders and from many citizens who were attempting to live normal lives in spite of the turmoil.

2. BACKGROUND TO THE CONFLICT

The conflict in Northern Ireland (Ulster) stems from the separation of Ulster from the rest of the island. The sea was less of a barrier than the mountains to the south of Ulster, so that there was a constant movement of people between Ulster and Scotland. Thus, even before the Plantation (settlement) at the beginning of the seventeenth century, the two eastern counties of Down and Antrim were already populated by farmers from Scotland.

Table 1. Overview of Conflict Analysis Studies

Author	Application	Date	Reference	Comments
Alexander & Saaty	Northern Ireland	1976, 1977	[1,3,4]	Stability of preferred outcome is tested; solution is shown to be robust.
Saaty & Gholamnezhad	Energy Conflict: U.S. vs. OPEC	1979, 1981	[6,9]	Conflict is modeled via game theory; AHP is used to obtain payoff matrix.
Tarbell & Saaty	South Africa	1980	[12]	Outcome stability is tested and possible coalitions are considered.
Alexander	Northern Ireland	1982	[2]	New parties, scenarios, and objectives are added.
Saaty, Vargas & Barzilay	Iran Hostage Rescue Operation	1982	[11]	AHP is used to explain disagreement between President Carter and his Secretary of State.
Saaty	Falkland Islands	1983	[7]	Hierarchies of benefits and costs are derived and ratios are used to reach a decision.
Vargas	Middle East	1983	[13]	A wide variety of objectives and scenarios are considered.
Saaty	Nuclear Balance: U.S. vs. U.S.S.R.	1983	[8]	Parity index concept for "bundles" of nuclear weapons held by U.S. and U.S.S.R. is developed.
Arbel & Novik	U.S. and Israel Relations	1985	[5]	Possible changes in nature of relationship are considered.

Thus, the people of Northern Ireland fall into two main groups. The majority (approximately two-thirds) descend from the Scots and English settlers of the early seventeenth century Plantation and of earlier migrations and are primarily Protestant. They wish to remain separate from the Republic of Ireland and to maintain the British connection. The minority (approximately one-third) are descendants of the Gaels and are primarily Roman Catholic. Some members of this group wish to unite Northern Ireland with the Republic of Ireland whose population is almost exclusively Roman Catholic.

When the Irish finally achieved independence from Great Britain in 1921, it was not quite what they wanted. Following the revised Government of Ireland Act, passed in 1920, the Northerners exercised the choice of establishing their own political entity of Northern Ireland. This new state would have its own Parliament, with certain powers, such as taxation and foreign policy, reserved to the British parliament. Subsequent governments of the Irish Free State, later the Republic of Ireland, refused to recognize this partition and claimed all of the island of Ireland. To help follow the key events in the history of Northern Ireland, we present a chronology in Figure 2.

At the outset, many Catholics refused to recognize the de jure existence of the state of Northern Ireland; they were thus regarded as disloyal. When it became clear that this new state was not a transient phenomenon, a growing number wanted to participate in policy making levels of government, but found that they were still regarded as potentially disloyal.

Meanwhile, there were those who wished to unite the two states and who were prepared to use violence to accomplish this objective. Chief among these groups was the Irish Republican Army (I.R.A.), a militant group based largely in the Republic of Ireland, which occasionally mounted campaigns of bombings and assassinations in Northern Ireland.

In 1969, the frustrations of many in the Catholic community led to increasing resentment on which the I.R.A. capitalized. The conflict in Northern Ireland escalated. Finally, in 1972, Britain suspended the Northern Ireland Parliament and government, and instituted direct rule from London through a Secretary of State. In 1973, a new Constitution for Northern Ireland was passed by the British Government, which established a Northern Ireland Assembly and Executive with very limited powers. In December, 1973, tripartite meetings (representatives from the British Government, the Northern Ireland Executive, and the government of the Republic of Ireland) led to the agreement to form a Council of Ireland which would consider problems of interest to both states in Ireland and ways in which the two states might be joined. Most important, such a Council would have legislative powers. This agreement was accompanied by a declaration, to which all parties subscribed, that there would be no further change in the constitutional position of Northern Ireland unless a majority in Northern Ireland agreed.

Figure 2. A Chronology of Key Events in Northern Ireland

Year	Event
1920	Government of Ireland Act passed
1921	Two states established in island: Northern Ireland and Irish Free State (Eire)
1949	Republic of Ireland established
1969	"Civil rights" riots in Northern Ireland
1972	Northern Ireland Parliament dissolved by British Government
	Direct rule from London instituted
1973	Northern Ireland Constitution passed
	Assembly and Executive established
	Sunningdale Conference: agreement to form Council of Ireland
1974	Constitutional Stoppage
	Northern Ireland Constitution suspended by British Government
	Direct rule from London reimposed
1975-76	Constitutional Convention for Northern Ireland
1982	Northern Ireland Assembly elected
1983	Northern Ireland Assembly dissolved
1985	Anglo-Irish Accords signed

This agreement aroused sharp resentment among the Protestant community, which rejected any possible connection with the Republic of Ireland. An immediate result was the repudiation of the Leader of the Northern Ireland Executive by his own party (February, 1974), followed shortly thereafter by a general strike,known as the Constitutional Stoppage. Supported by almost all Protestants, this strike brought all activity to a standstill for almost a month and led to the resignation of the Executive. Since then, Northern Ireland has been ruled directly by the British Government through a Secretary of State for Northern Ireland.

In 1975-76, a Constitutional Convention in Northern Ireland recommended the restoration of Parliamentary government and the establishment of legislative committees based on the United States model. An important feature was that many important chairmanships would be held by members of minority parties. This report was rejected by the British Government.

Successive Secretaries of State have attempted to create new political structures, but with no success. There have been attempts by the British Government to negotiate with the I.R.A. about the future of Northern Ireland, but this has served only to inflame the non-violent members of both communities.

In 1982, another Assembly was elected by the people of Northern Ireland. This body had only scrutinizing and advisory functions and had neither executive nor legislative functions. The Assembly was charged with developing a framework for gradual devolution of some power to the Northern Ireland people. The constraints placed by the British on possible proposals to be considered by the Parliament in London (70% acceptance of such proposals in the Assembly) made progress impossible. The Assembly was dissolved.

In October, 1985, the Anglo Irish Accords between Great Britain and the Republic of Ireland were signed. These gave the Republic of Ireland considerable input into all aspects of life in Northern Ireland, through meetings of members of both governments and through a joint Secretariat (based in Northern Ireland) of civil servants from both countries. This has caused widespread and increasing resentment in the majority (Protestant) community in Northern Ireland. Meanwhile, the I.R.A. has stepped up the violence, and bombings and killings occur every day. The potential for all-out civil war is at its peak.

This brief account does not discuss such topics as the allegations of discrimination, the different factions of the I.R.A. and of related groups, and the different political parties. In particular, it should be noted that the use of the terms "Catholic" and "Protestant" is for convenience only; this is not a religious conflict, as will be noted in the objectives of the parties. The denominational identification is made because this is the way in which the world at large views the problem.

3. CONSTRUCTING THE CONFLICT HIERARCHY FOR NORTHERN IRELAND

Following the format of the general hierarchy depicted in Figure 1, we develop a hierarchy for the current Northern Ireland conflict. We start by constructing the first level, i.e., parties to the conflict. The complete hierarchy is shown in Figure 3.

3.1 Parties to the Conflict

At this level of the hierarchy, we must identify those individuals or groups who may have an influence on the outcome.

Figure 3. Conflict Hierarchy for Northern Ireland

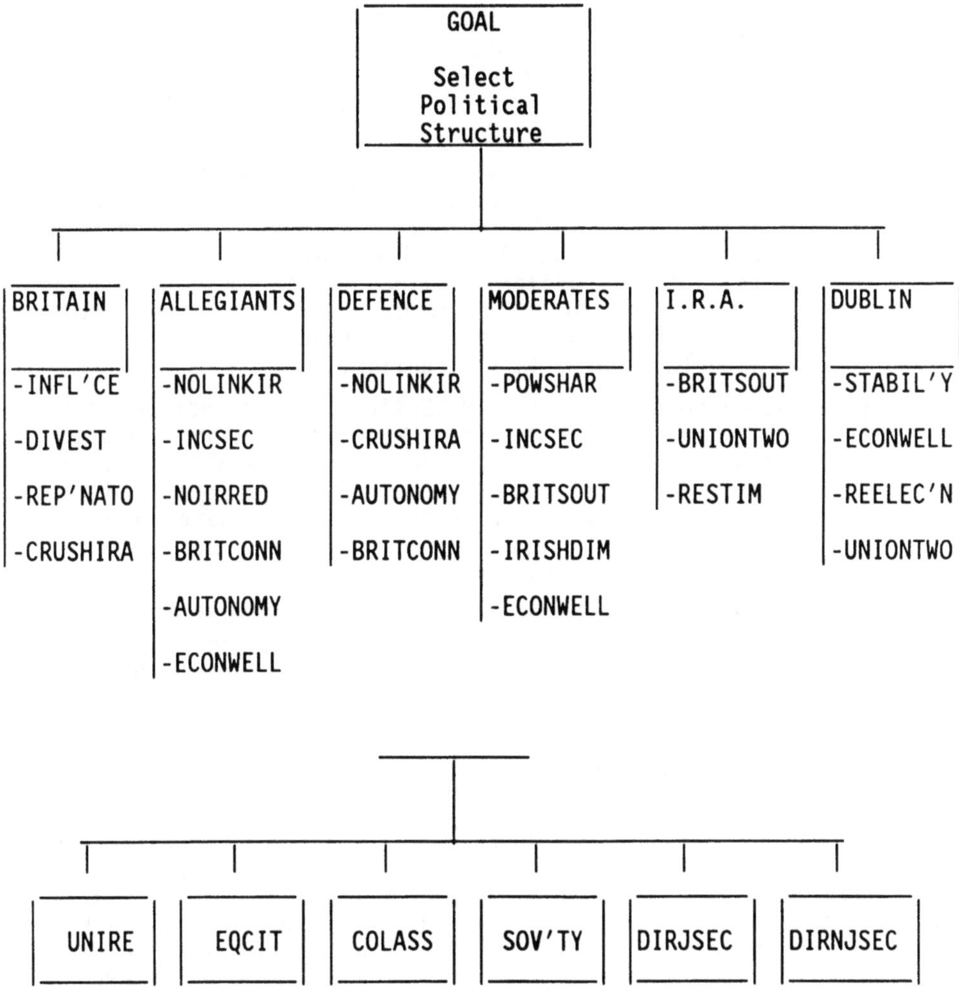

As in the 1982 study, these parties are:

(a) The British Government (BRITAIN), which controls Northern Ireland.

(b) The Protestant community (ALLEGIANTS), often known as loyalists, which wants Northern Ireland to remain separate from the Republic of Ireland and which would find a substantial measure of minority participation in government acceptable.

(c) The Protestant paramilitary defence forces (DEFENCE). Some of the most interesting political initiatives and position papers in Northern Ireland have come from study groups and political parties set up by the main Protestant paramilitary forces. Since these groups are in a position to deploy force against any attempts to overthrow the state of Northern Ireland, they constitute an important party in the conflict.

(d) The Catholic community (MODERATES), which includes those who would prefer to join Northern Ireland with the Republic of Ireland and also those who would be content to have Northern Ireland remain separate, provided that a structure which provides for adequate minority participation is established. (This group does not include those who support violence.)

(e) The Irish Republican Army (I.R.A.), including both the Provisional and Official subgroups, as well as their supporters, which uses violence in an attempt to create a united Ireland.

(f) The Government of the Republic of Ireland (DUBLIN), which makes formal claim in its constitution to the territory of Northern Ireland.

Other parties could be added to this set. There have been indications that the United States is no longer adopting a "hands-off" policy in relation to the conflict; for example, official representatives of the executive branch have publicly expressed a desire for a united Ireland. It is not clear whether this reflects the intense lobbying efforts of Irish irredentist groups, such as the Irish National Caucus and Irish Northern Aid, the influence of Irish-Americans in the White House, or the possibility of using Northern Ireland as a bargaining chip in bringing the Republic of Ireland into NATO. The United States also makes annual grants to a fund designed to support the Anglo-Irish Accords. However, since the United States has not officially changed its position of non-interference in the affairs of Northern Ireland and has taken no direct actions in the dispute, it is not defined explicitly as a party to the conflict; its influence is represented indirectly by increases in the power of BRITAIN and of DUBLIN.

3.2 The Objectives of the Parties

3.2.1 Objectives of BRITAIN

Britain seeks to maintain its sphere of influence and is concerned that any political solution should not diminish this influence in either Northern Ireland or the Republic of Ireland. At the same time, the British public has become weary of the Ulster problem and many would like to see Britain withdraw its soldiers. Further, the pressure of I.R.A. propaganda overseas, notably in the United States, has made many unsympathetic to Britain's role in Ulster. For these and other reasons, Britain would like to divest itself of the Ulster problem.

Many in Ulster think that Britain would like to help the United States add the Republic of Ireland to NATO and is prepared to use Northern Ireland as a bargaining chip to accomplish this. In fact, it is highly unlikely that this aim could be achieved: the tradition of neutrality in the Republic of Ireland is unusually strong and is shared by many politicians in both parties. However, it is possible that the British Government is unaware of this.

In recent years, the activities of the I.R.A. have extended to mainland Britain, so that the British government is now adopting a harder line towards terrorism. Thus, defeating the I.R.A. has become an objective.

To summarize, the objectives of BRITAIN are:

INFL'CE	maintain sphere of influence
DIVEST	divest of Ulster problem
REP'NATO	add the Republic of Ireland to NATO
CRUSHIRA	defeat the Irish Republican Army.

3.2.2 Objectives of the ALLEGIANTS

Above all other considerations, the ALLEGIANTS (the Protestant majority) are concerned that Northern Ireland should not be annexed by the Republic of Ireland. They reject any link with the Republic of Ireland and do not want any Irish irredentists in government. They seek a substantial measure of political autonomy, but would like to maintain the British connection.

The level of violence has become totally unacceptable. Security in Northern Ireland has worsened. There have been many murders by the I.R.A. and the British Army has not been in full control. For the citizens of Northern Ireland, security is fundamentally important.

There is also a concern for economic well-being. There has been much industrial development, although the level of unemployment in western areas is unacceptably high.

Thus, we may summarize the objectives of the ALLEGIANTS as follows:

NOLINKIR	no link with the Republic of Ireland
INCSEC	increased security
NOIRRED	no Irish irredentists in government
BRITCONN	maintain British connection
AUTONOMY	political autonomy
ECONWELL	economic well-being.

3.2.3 Objectives of DEFENCE

The Protestant paramilitary forces have many objectives in common with the ALLEGIANTS, although the relative importance of those objectives may be different. For example, they do not want a link with the Republic of Ireland, and they want, if possible, both political autonomy and the British connection. The major paramilitary organizations were created from local defense groups which were formed when many in the majority community decided that the British were not protecting them from the Irish Republican Army. Thus, a primary concern is the defeat of the I.R.A.

Thus, the objectives of the Protestant DEFENCE forces are:

NOLINKIR	no link with the Republic of Ireland
CRUSHIRA	defeat the Irish Republic Army
AUTONOMY	political autonomy
BRITCONN	maintain British connection.

3.2.4 Objectives of the MODERATES

The MODERATES (the Roman Catholic minority, excluding those who support violence) seek a share in political power and increased security. Many want the British to withdraw from Northern Ireland. Because of their Irish nationalism, they seek some form of "Irish dimension" in any political solution. They are also concerned with economic well-being.

Thus, the objectives of the MODERATES are:

POWSHAR	share in political power
INCSEC	increased security
BRITSOUT	British withdrawal
IRISHDIM	Irish dimension in political structure
ECONWELL	economic well-being.

3.2.5 Objectives of the I.R.A.

The I.R.A. is committed to driving out the British and to creating a United Ireland. The failure of a number of their plans has also led to a desire to restore their image.

Thus, the objectives of the I.R.A. are:

BRITSOUT	British withdrawal
UNIONTWO	unite the two countries (Northern Ireland and the Republic of Ireland)
RESTIM	restore own image.

3.2.6 Objectives of DUBLIN

A major concern of the government of the Republic of Ireland (DUBLIN) is that a political solution of the Ulster problem might affect the stability of the Republic of Ireland. They also wish to maintain healthy economic conditions. Both major political parties have the long-term objective of creating a United Ireland. The party in power is also concerned with re-election.

To summarize, the objectives of the Republic of Ireland are:

STABIL'Y	maintain stability
ECONWELL	economic well-being (of Republic)
REELEC'N	re-election of party in power in Republic
UNIONTWO	unite two countries (Northern Ireland and Republic of Ireland).

3.3 Political Structures

We will use the four basic political structures defined in the 1982 analysis and, based upon discussions with knowledgeable Ulster friends, we will include two new structures: one that represents the status quo and another for an "almost" status quo. These structures are listed below.

(a) UNITED IRELAND (UNIRE). All possible ways in which the two states in the island might be joined, including such concepts as federation and confederation.

(b) EQUAL CITIZENSHIP (EQCIT). Full integration of Northern Ireland into the United Kingdom. Northern Ireland sends members to the British Parliament but they are neither consulted nor informed on matters concerning Northern Ireland, and the British political parties do not organize in Northern Ireland.

(c) COLONIAL ASSEMBLY (COLASS). An elected representative body in Northern Ireland, but one subordinate to the British Government. The range of powers of such a body could vary from the extremely limited powers of the short-lived Assembly of 1973-74 to those of the more powerful regional Parliament which was prorogued (suspended) in 1972.

(d) SOVEREIGNTY (SOV'TY). Northern Ireland would have an independent government which would not be subordinate either to Britain or to the Republic of Ireland. This could include a Dominion or a Republic within the

British Commonwealth, or a constituent state of the European Community.

(e) DIRECT RULE WITH JOINT SECRETARIAT (DIRJSEC). The status quo, where Britain rules Northern Ireland directly. The joint council of Britain and the Republic of Ireland discusses policies for Northern Ireland, and the Joint Secretariat of British and Irish civil servants is involved in the administration of the country.

(f) DIRECT RULE WITHOUT JOINT SECRETARIAT (DIRNJSEC). As in structure (e) above, but with no involvement of the Republic of Ireland at either policy making or administrative levels.

4. THE JUDGMENTS AND CALCULATIONS

We now develop the pairwise comparison matrices at each level of the hierarchy. These judgments are made by the author (who has visited Northern Ireland many times over the last several years, most recently in the summer of 1988), but reflect many long hours of discussion with political leaders, academics, paramilitary and community leaders, and with concerned citizens of Northern Ireland, all of whom expressed themselves very freely. There is a growing awareness in all sections of the community that any solution must involve both majority and minority communities. Some preferences, such as those of the I.R.A., had to be assessed from their actions and, to a lesser extent, from their statements. The author has strong ties of kinship with the Protestant community (ALLEGIANTS); every effort has been made not to allow this to color the analysis.

At the first level of the hierarchy, we derive the matrix of comparisons by asking such questions as: Does BRITAIN have more power than DEFENCE to influence final outcome? If so, how much more powerful is BRITAIN? The answer here lies between "equality" (a value of 1) and "moderately more powerful" (a value of 3); thus, this entry is 2. The matrix of comparisons and the weights of each party are given in Table 2.

At the remaining levels of the hierarchy, we develop similar pairwise comparison matrices (these are available from the author upon request) and we use hierarchic composition to produce the set of weights for the alternatives that is shown in Table 3.

SOVEREIGNTY has by far the largest weight; EQUAL CITIZENSHIP and COLONIAL ASSEMBLY, almost equal, come in a poor second. The consistency ratio of the entire hierarchy is 0.02. Thus, some form of independent structure, the same result as in the 1976 and 1982 studies, most satisfies the needs and objectives of the parties.

Table 2. Matrix of Comparisons of Power of Parties to Conflict

	BRITAIN	ALLEGIANTS	DEFENCE	MODERATES	IRA	DUBLIN
BRITAIN	1	2	2	4	9	9
ALLEGIANTS	1/2	1	1	2	6	7
DEFENCE	1/2	1	1	2	6	7
MODERATES	1/4	1/2	1/2	1	3	4
IRA	1/9	1/6	1/6	1/3	1	1
DUBLIN	1/9	1/7	1/7	1/4	1	1
WEIGHTS	0.386	0.216	0.216	0.111	0.037	0.034

Table 3. Final Set of Weights

Political Structure	Weight
UNITED IRELAND	.143
EQUAL CITIZENSHIP	.170
COLONIAL ASSEMBLY	.167
SOVEREIGNTY	.291
DIRECT RULE, JOINT SECRETARIAT	.120
DIRECT RULE, NO JOINT SECRETARIAT	.110

5. THE BACKWARD PROCESS AND REPEATED FORWARD PROCESS

The process used thus far in this analysis has answered the question: given the present actors and their current objectives, which outcome is most likely to emerge? This descriptive approach is known as the **forward process**. An alternative question is: given a desired future outcome, what can be done to achieve this outcome? Here one works backwards to see what changes might be made to produce a desired outcome. This normative approach is known as the **backward process**.

A combination of such approaches is often used in analysis and planning (see [3,4]). It should be noted that the term "desired outcome" does not imply any preference on the part of the analyst. It may refer to the preference of one party to the conflict who does not like the resultant outcome from the forward process and who wishes to see what changes would be necessary to alter the result.

From this analysis, the preferences for each party are clear. The ALLEGIANTS and DEFENCE both prefer SOVEREIGNTY; BRITAIN and the I.R.A. prefer UNITED IRELAND; the MODERATES prefer COLONIAL ASSEMBLY; DUBLIN prefers DIRECT RULE WITH JOINT SECRETARIAT.

Those parties for whom SOVEREIGNTY is not the outcome of choice may desire to see if this result could be altered by changes in the hierarchy. In essence, we study the stability of the SOVEREIGNTY outcome. Since it is the ALLEGIANTS and DEFENCE who prefer SOVEREIGNTY, we may test the stability of this outcome by varying the power of these parties and distributing the removed power to the other parties. We then repeat the forward process and examine the new outcomes.

First, we remove ALLEGIANTS and DEFENCE completely from the analysis; this removes their combined power of .432, and leaves .568 for the remaining parties. The power of the remaining parties is then divided by .568 to give the following weights: BRITAIN (.680), MODERATES (.195), I.R.A. (.065), and DUBLIN (.060). The forward process can be repeated using these weights. If all else is unchanged, the net effect of this is to divide all their contributory weights to the political structures by .568. This process yields the set of weights shown in Table 4.

Table 4. Forward Process Weights

Political Structure	Weight
UNITED IRELAND	.232
EQUAL CITIZENSHIP	.103
COLONIAL ASSEMBLY	.127
SOVEREIGNTY	.229
DIRECT RULE, JOINT SECRETARIAT	.181
DIRECT RULE, NO JOINT SECRETARIAT	.131

The weight for UNITED IRELAND is now slightly larger than the weight for SOVEREIGNTY. The SOVEREIGNTY outcome still has a high weight because, while it is not the first choice of the other parties, it meets many of their needs.

This result suggests that a very low level of power for ALLEGIANTS and DEFENCE would still give SOVEREIGNTY the greatest weight. The weights for these two parties were then varied from .01 to .1 of their original weights. When their weights are set at .025 of the original level, and the remaining power is distributed to the other parties, the weights for UNITED IRELAND and SOVEREIGNTY are equal. (Multiply the contributory weights for ALLEGIANTS and for DEFENCE by .025 and multiply the contributory weights for the remaining parties by [1 - .025(.432)]/[1 - .432] = 1/.574). If the weights for ALLEGIANTS and DEFENCE are higher than .025 of the original level, SOVEREIGNTY has the highest weight.

Thus, this level of power provides a **threshold of power** for these two parties. It should be noted that, since their power is reduced to .025 of the original, while at the same time the power of every other party is divided by .574, the power of these parties, related to the others, is factored by .025 x .574 to reach the break-even point; i.e., the relative power of the two parties has to be reduced to approximately one-seventieth of the original in order to change the outcome.

An alternative approach to testing the stability of SOVEREIGNTY would be to add an objective of weakening the majority parties to the objectives of the remaining parties. However, it is obvious that this objective would be resisted and that the most important factor is power to influence the outcome. It should also be noted that the I.R.A. objective of British withdrawal includes the disarming of both DEFENCE and ALLEGIANTS.

These results are consistent with recent governmental policies. There have been strong efforts to weaken the two groups from the majority community, presumably to leave Northern Ireland vulnerable to annexation by the Republic of Ireland. We conclude that the SOVEREIGNTY outcome is unusually stable.

We now compare the results of the current analysis with earlier results in Table 5. In the 1976 study, we considered two ASSEMBLY outcomes--Assembly without Council of Ireland and Assembly with Council of Ireland. Similarly, two DOMINION structures were also considered. We combine the two ASSEMBLY structures as COLONIAL ASSEMBLY and we combine the two DOMINION structures as SOVEREIGNTY in Table 5.

In the 1982 study, the two ASSEMBLY outcomes were combined, and the two DOMINION outcomes were replaced by Independence. In the 1988 study, Integrated Parliament is replaced by EQUAL CITIZENSHIP, and Independence by SOVEREIGNTY. The DIRECT RULE options were not considered in either 1976 or 1982; at that time, no one thought it possible that the political process could remain in abeyance for any appreciable length of time. In order to make direct comparisons, we remove the DIRECT RULE outcomes from the 1988 study and distribute their weights to the other structures; this gives a modified set for 1988.

We see that the weight for SOVEREIGNTY in 1988* is higher than in 1982 but less than in 1976. The weights for UNITED IRELAND and for EQUAL CITIZENSHIP have grown over the period, and the weight for COLONIAL ASSEMBLY increased from 1976 to 1982 but decreased sharply from 1982 to 1988 and 1988* to approximately the same level as for EQUAL CITIZENSHIP. However, in each year, the overwhelming preference is for an outcome which does not involve a political link with the Republic of Ireland; this agrees with all recent election results. (The overwhelming majority of the votes have gone to parties and to candidates who oppose a United Ireland or any link with the Republic of Ireland.)

Table 5. Changes in Political Structures from 1976 to 1988

Political Structure	Weight			
	1976	1982	1988	1988*
UNITED IRELAND	.148	.157	.143	.186
EQUAL CITIZENSHIP	.154	.193	.170	.221
COLONIAL ASSEMBLY	.289	.308	.167	.217
SOVEREIGNTY	.409	.343	.291	.378
DIRECT RULE, JOINT SECRETARIAT	NC	NC	.120	NC
DIRECT RULE, NO JOINT SECRETARIAT	NC	NC	.110	NC

NC Not Considered
* Modified Results

6. CONCLUSIONS

The use of the AHP has again provided valuable insights into the structure of a conflict problem and has suggested a solution for the conflict in Northern Ireland. As in the previous studies, the outcome which best satisfies the needs of all parties is some form of legislative independence, with Northern Ireland subordinate neither to the British nor to the Irish Government. This outcome could take such practical forms as a Dominion or Republic within the British Commonwealth or a constituent state of the European Community. The essential concern seems to be that the people of Northern Ireland should be free to determine their own form of government.

7. REFERENCES

1. J. Alexander, "A Study of Conflict in Northern Ireland: An Application of Metagame Theory," **Journal of Peace Science, 2,** 113-133 (1976).

2. J. Alexander, "Priorities and Preferences in Conflict Resolution," **Mathematics and Computers in Simulation, 25,** 108-119 (1982).

3. J. Alexander and T. Saaty, "The Forward and Backward Processes of Conflict Analysis," **Behavioral Science, 22,** 87-98 (1977).

4. J. Alexander and T. Saaty, "Stability Analysis of the Forward-Backward Process: Northern Ireland Case Study," **Behavioral Science, 22,** 375-382 (1977).

5. A. Arbel and N. Novik, "U.S. Pressure on Israel--Likelihood and Scope," **Journal of Conflict Resolution, 29,** 253-282 (1985).

6. A. Gholamnezhad, "Critical Choices for OPEC Members and the United States," **Journal of Conflict Resolution, 25,** 115-143 (1981).

7. T. Saaty, "Conflict Resolution and the Falkland Islands Invasions," **Interfaces, 13 (6),** 68-83 (1983).

8. T. Saaty, "Impact of Disarmament Nuclear Package Reductions," in **Quantitative Assessments in Arms Control**, R.Avenhaus and R. Huser, editors, Plenum Press, New York, New York (1983).

9. T. Saaty, "The US-OPEC Energy Conflict--The Payoff Matrix and the Analytic Hierarchy Process," **International Journal of Game Theory, 8,** 225-234 (1979).

10. T. Saaty and J. Alexander, **A New Logic to Resolve Conflicts**, forthcoming in 1989.

11. T. Saaty, L. Vargas, and A. Barzilay, "High-level Decisions: A Lesson From the Iran Hostage Rescue Operation," **Decision Sciences, 13,** 185-206 (1982).

12. D. Tarbell and T. Saaty, "The Conflict in South Africa: Directed or Chaotic," **Journal of Peace Science, 4,** 151-168 (1980).

13. L. Vargas, "Prospects for the Middle East: Is a Peaceful Settlement Attainable?" **European Journal of Operations Research, 14,** 169-192 (1983).

SITE SELECTION FOR A LARGE SCALE INTEGRATED CIRCUITS FACTORY

Kaoru Tone
Graduate School of Policy Sciences
Saitama University
Urawa, Saitama 338, Japan

Shigeru Yanagisawa
General Manager, Production Control Division
OKI Electric Industry Company, Limited
550-1 Higashiasakawacho
Hachioji-Shi
Tokyo 193, Japan

ABSTRACT

A Japanese manufacturer of electronic goods is planning to construct a new, state-of-the-art factory to manufacture large scale integrated circuits. A team of company managers must decide on the appropriate location for the factory taking many different criteria into account. This paper reports on the decision process that the project team employed to model the site selection problem.

1. BACKGROUND OF THE COMPANY

In 1881, OKI Electric Industry Company, Ltd. began operations as the first manufacturer of telephones in Japan. Today, it is one of the world leaders in the development, manufacture, and sales of telecommunication systems, information processing systems, and electronic devices. OKI has nine main production factories in Japan (located in the cities of Tokyo (3), Saitama (2), Yamanashi (1), Gunma (2), and Shizuoka (1)) as well as overseas production centers (located in the United States and Scotland) and is technically affiliated with IBM and AT&T. As of March 1988, OKI employed 13,800 workers.

In fiscal 1987, OKI was the 15th largest electric company in Japan with sales of 420 billion yen. Thirty percent of sales came from telecommunications systems, 45 percent from information processing systems, and 23 percent from electronic devices. We point out that more than 80 percent of the electronic devices sales are derived from the Metal Oxide Semiconductor Integrated Circuit (MOS IC). In 1987, OKI's worldwide market share was 5th in IC, 11th in MOS IC, 14th in MOS Logic, and 16th in MOS Application Specific Integrated Circuits, according to a 1988 issue of the **Dataquest Newsletter**. The Electronic Devices Group, which oversees integrated circuit production, has 7,000 employees including affiliates.

The main production centers for integrated circuits are located in Hachioji (Tokyo) and Miyazaki (Kyushu). Each of the centers contains two factories. However, in 1985, customer demands for integrated circuits were so large that the company could not fill all orders. Also, with the coming submicron age, the company foresaw the need to build a new, leading-edge factory.

In the remainder of this paper, we focus on the decision-making process used by OKI to select an appropriate site for a new factory that would manufacture large scale integrated circuits.

2. OUTLINE OF THE PROJECT

In July 1985, a project team was selected from the Electronic Devices Group and was charged with the task of proposing a site selection and construction plan for a new leading-edge factory. The project team reported directly to the Board of Directors and consisted of the following members:

. **Group A** -- 5 managers from the integrated circuits division with experience in manufacturing, technology, accounting, and general affairs,

. **Group B** -- 5 experts in integrated circuits with expertise in design, processing, and infrastructure technology, and

. **Group C** -- 5 leaders in factory management and manufacturing technology that will be in charge of the new, completed factory.

The manager of the integrated circuits division headed the project team.

The objectives of the project team were (1) to clearly define a next generation, state-of-the-art IC factory that is capable of meeting demand and that is also suitable for the submicron age, (2) to select an appropriate site, and (3) to report their results to the Board of Directors within two years.

The first step in the decision-making process required the project team to discuss the overall plan of the new factory and to gather information relevant to the project. For example, the team needed to decide upon (1) the types of integrated circuits to be produced at the new factory, (2) the production processes that would be required to produce the ICs, (3) a suitable infrastructure for the processes, and (4) a desirable working environment at the factory.

When details about the project were "leaked" to the public, many cities became quite interested in obtaining the factory. Several cities were very quick to recommend themselves as the appropriate site: They sent representatives to discuss the site selection with OKI's management and project team. In all, more than twenty sites from around the world were identified as candidate locations for the new factory.

After the potential sites were identified, the project team screened this initial set on three key decision factors and eliminated several sites. These sites were dropped from further consideration since they lacked (1) a water supply sufficient for the production of integrated circuits, (2) a large labor pool to operate the factory (about 1,800 employees would be hired), and (3) sufficient land for future expansion (more than 20 hectares would be necessary).

To screen the remaining candidates, the project team gathered more data about the sites and augmented and modified a checklist of decision factors that had been used in a previous site selection decision. (A shorter list was used to help select the OKI site at Miyazaki.) A new checklist of over 120 key items was compiled on which to compare the potential sites. A partial list of factors and the team's evaluation of three sites are shown in Table 1. From this table, we see that the three decision factors -- technical, economic, and social -- are broken down into important items and each item is further decomposed into subitems. For example, the team judges the education and welfare capabilities of each site by examining the region's schools, educational standards, medical services, and hospitals. Each candidate site is then evaluated on a four-value scale-- excellent, fair, less than fair, and poor -- with respect to its performance on each subitem. The checklist approach helped the project team narrow down the candidate list to three sites: **Site A** (a town in the Kinki area of central Japan), **Site B** (a town on Shikoku Island in southern Japan), and **Site C** (a town in the Tohoku area of northern Japan). The final step in the decision-making process called for the project team to compare Sites A, B, and C in detail, to select the best site, and then to make a recommendation to the Board of Directors. However, the team found it difficult to make a selection based only on the checklist approach. The three sites exhibited the same score on many of the subitems and the team had no way of weighting the factors, items, and subitems listed in Table 1. For example, the attitudes and judgments of three groups represented on the project team widely differed on the importance of the technical, economic, and social factors. As a result, it became very difficult for the team to reach a consensus on the best site for the new factory.

At this step in the process, a group of three managers (one from Groups A, B, and C) was organized and charged with the task of ranking the three candidates. They decided to apply the Analytic Hierarchy Process to help model their decision problem. The three managers believed that the AHP was the appropriate method since it potentially could provide a rational way of ranking sites when conflicts exist in weighting key decision factors.

3. THE DECISION HIERARCHY

To construct the decision hierarchy for the site selection problem, the three managers excluded the subitems listed in Table 1 for which the three sites (A, B, and C) had the same rating. For example, the sites received identical evaluations on the two tax subitems (i.e., A, B, and C each received a fair rating on tax and exemption); therefore, the tax item is not included in the hierarchy. The final decision hierarchy is shown in Figure 1.

Table 1. Checklist for Site Selection

Factor	Item	Subitem	A	B	C
technical	situation of site	base	e	e	e
		area	e	f	e
		environments	f	f	e
		approach roads	f	f	e
	water supply	amount	e	e	e
		quality	e	e	e
	electricity	power	e	e	e
		break down	l	l	l
	supporting industries	gas	e	e	e
		chemicals	e	f	f
		materials	e	f	e
		maintenance	e	f	f
	weather	rain	f	f	f
		snow fall	e	e	l
		earthquake	f	e	l
		volcanic eruption	f	f	f
		injury from salt	l	l	f
economic	initial investment	real estate	e	f	f
		construction cost	f	f	f
		water supply (city/well water)	e	e	e
		electric transmission	f	f	e
	energy cost	water	e	e	e
		electricity	l	l	l
		material	f	f	f
	traffic and transportation	technical staff	e	e	f
		business staff	e	e	f
		production	e	f	f
	tax	tax	f	f	f
		exemption	f	f	f
	labor	recruit	e	e	f
		wage rate	e	f	e
		quality	e	e	e
		university graduate	f	l	e
		union	e	e	e
	public agencies	posture to industries	e	e	e
		distance to public agencies	f	f	f

e:excellent, f:fair, l:less than fair, p:poor

Table 1. (continued)

social	legal regulation	factory location act.	e e e
		building std. act.	e e e
		environmental pollution prevention act.	f f f
		fire service act.	f f f
	regional characteristics	living expenses	e f e
		shift system	e e e
		rate of absenteeism	f e f
		transference	e f l
		productivity	f f f
	housing	company's housing	f f f
		employee's housing	f l e
		distance from residential district	e e e
	education/ welfare	school	e f f
		educational standard	f f e
		medical services	f e f
		hospitals	e e f
	culture/ shopping	cultural facilities	f f f
		shops	e e e
		police stations	e e e
		fire houses	f f f
		restaurants	e e e

To generate the pairwise comparison matrices at each level, the group considered three different scenarios: (1) new technology is judged the most important (Case 1), (2) strategic management is rated highest (Case 2), and (3) personnel management is the most important (Case 3). Under each scenario, the group judged 'engineer' the most important element at the second level. The weights for this level are given in Table 2. At each level of the hierarchy, the comparisons were generated by the group of three managers with help from the entire project team. We point out that at the third level of the hierarchy, the technical factor always received the largest weight. These weights are given in Table 3. The remaining matrices were also generated taking the three scenarios into account.

Table 2. Weights at the Second Level

	Case 1	Case 2	Case 3
Engineer	.64	.43	.46
Manager	.11	.43	.09
New Factory Personnel	.26	.14	.46

Figure 1. Hierarchy Structure

```
Level 1     2           3            4             5                6

site-----|-engineer-|-technical-|-site-------|-area----------|
selection|          |           |            |-environment   |
         |-manager--|           |            |-approach------|
         |          |           |            |               |
         |          |           |-supporting-|-chemicals-----|
         |-new------|           | industries |-materials-----|
           factory  |           |            |-maintenance---|
           personnel|           |            |               |
                    |           |-weather----|-snow fall-----|
                    |                        |-earthquake----|
                    |                        |-injury from---|
                    |                           salt         |
                    |                                        |
                    |-economic--|-initial----|-real estate---|   -site A
                    |           | investment |-elec.trans.---|
                    |           |            |               |
                    |           |-traffic----|-tech.staff----|
                    |           | /transp.   |-bus. staff----|   -site B
                    |           |            |-production----|
                    |           |            |               |
                    |           |-labor------|-recruit-------|
                    |                        |-wage rate-----|   -site C
                    |                        |-university----|
                    |                                        |
                    |-social----|-regional---|-living--------|
                                |            |-absenteeism---|
                                |            |-transference--|
                                |            |               |
                                |-housing----|-employee------|
                                |            |               |
                                |-education--|-school--------|
                                | /welfare   |-standard------|
                                             |-medical-------|
                                             |-hospitals-----|
```

Table 3. Weights at the Third Level

	Case 1	Case 2	Case 3
Technical	.73	.63	.64
Economic	.19	.24	.11
Social	.08	.14	.26

Table 4 displays the overall priorities of the three candidate sites. Although Site C has the top priority in every case, its superiority over Site A is too small. It is quite difficult to select Site C as the final choice without further thought. Figure 2 shows the decision hierarchy in Case 1 with the corresponding weights attached to each factor. From this we see that the items with the highest scores on each level are (1) engineer and new factory personnel on the second level, (2) technical factor on the third level, (3) situation of site and supporting industries on the fourth level, and (4) area, maintenance, environments, recruit, living expenses, and earthquake on the fifth level.

Table 4. Overall Priorities

Sites	Case 1	Case 2	Case 3
A	.39	.40	.39
B	.19	.19	.19
C	.41	.41	.42

Table 5 shows the differences between Sites A and C with respect to the highly scored items on Level 5. Site A is superior to Site C in maintenance, recruit, and earthquake and Site C is superior to Site A in area, environments, and living expenses.

Table 5. Priorities of Sites A and C on Level 5 Items

Sites	Area	Maintenance	Environments	Recruit	Living Expenses	Earthquake
A	.07	.11	.02	.05	.01	.01
C	.18	.01	.09	.01	.03	.00

We can also examine the three sites with respect to the Level 2 actors. These weights are displayed in Table 6. From this table we see that the scores of Site B for every actor are inferior to those of Sites A and C. Therefore, under any scenario or any weights of actors, Site B is inferior to Sites A and C.

Figure 2. AHP Results (Case 1)

Level 1	2	3	4	5		6
site selection	engineer .64	technical .70	site .44	area	.29	
	manager .11			environment	.12	
	new factory personnel .26			approach	.03	
			supporting industries .18	chemicals	.03	
				materials	.01	
				maintenance	.14	
			weather .07	snow fall	.01	
				earthquake	.05	
				injury from salt	.02	
		economic .17	initial investment .04	real estate	.04	
				electric transmission	.01	site A .39
			traffic /transp. .02	tech. staff	.01	
				bus. staff	.00	
				production	.01	site B .19
			labor .11	recruit	.08	
				wage rate	.02	
				university	.01	
		social .13	regional .08	living	.05	site C .41
				absenteeism	.01	
				transference	.02	
			housing .01	employee	.01	
			education /welfare .04	school	.02	
				standard	.01	
				medical	.00	
				hospitals	.01	

Table 6. Weights of Sites A, B, and C With Respect to the Level 2 Actors

	Site A	Site B	Site C
Engineer	.397	.188	.414
Manager	.407	.196	.396
New Factory Personnel	.379	.199	.423

4. CONCLUSIONS

Using the results from the Analytic Hierarchy Process, the project team concluded that Site B should be dropped from further consideration. Although Site C scored slightly better than Site A in all cases, the difference was too small to make a decisive conclusion. The manager of the project team reported these results to the Board of Directors and after deliberation the Board decided to locate the new factory at Site C. In particular, two additional factors prompted the Board's decision: (1) Site C can cooperate with nearby Tohoku University which is well known for its integrated circuit research and (2) there is a strong labor pool of IC researchers in the area near Site C.

In 1987, OKI decided to build a state-of-the-art integrated circuit factory in the Tohuku area of northern Japan (i.e., Site C). The new factory would include silicon wafer fabrication and assembly capabilities. The factory is currently under construction and is scheduled to come on line by the end of 1988.

To summarize, the project team could not differentiate between Sites A, B, and C based on the checklist approach. In fact, this approach seems to imply that "Site A is better than Site C" (A received 32 excellent scores and C received 27 excellent scores in Table 1). The simplistic checklist approach could have provided the "wrong" recommendation. However, the results of the Analytic Hierarchy Process indicated otherwise. The AHP results were very persuasive in helping OKI reach a decision.

BUSINESS STRATEGY FORMULATION FOR A FINANCIAL INSTITUTION IN A DEVELOPING COUNTRY

Luis G. Vargas and J. Bernat Roura-Agusti
Joseph M. Katz Graduate School of Business
University of Pittsburgh
Pittsburgh, Pennsylvania 15260

ABSTRACT

This paper provides a summary of a project done in 1986 for an organization in a Central American country. The purpose of the work was to develop a global strategy and some functional strategies to improve the organization's image and to generate new sources of funds.

1. PROJECT BACKGROUND

FDC is a government agency of a Central American country that coordinates fifty Savings and Loans Associations (SLAs) which are comprised of approximately 87,000 members. The actual name of the organization has been changed to FDC to maintain confidentiality. The organization was created in 1943 to help small businesses which were left unprotected after the banking reform that took place in the 1930s. FDC presently acts as the coordinator of a cooperative formed by the SLAs in 1938. This cooperative movement was the product of a plan developed by the General Assembly of Coffee Producers in their attempt to help small farmers. The government greatly influences the structure of FDC through the appointment of its CEO by the President of the Republic. The organizational structure of FDC is given in Figure 1.

The main objectives of FDC are to:

- promote the cooperative movement and, in particular, to help small businesses obtain funds,

- foster education (through continuing support of seminars for farmers and small entrepreneurs on agricultural and industrial developments),

- participate and collaborate with societies and institutions related to the cooperative movement,

- authorize the formation and functioning of SLAs, and audit them, and

- serve as the intermediary between the SLAs and third parties (e.g., the Federal Reserve Bank of the Republic and international agencies).

FDC allocates resources from three different funds--the FDN fund, the EMS fund, and the PTN fund. The first fund consists of five departments: Agriculture, Normal Operations, Rural Development, Urban Development, and Popular Credit. The main objective of these departments is to provide loans to individuals or companies with limited or no means of obtaining credit

Figure 1. The Organizational Structure of FDC

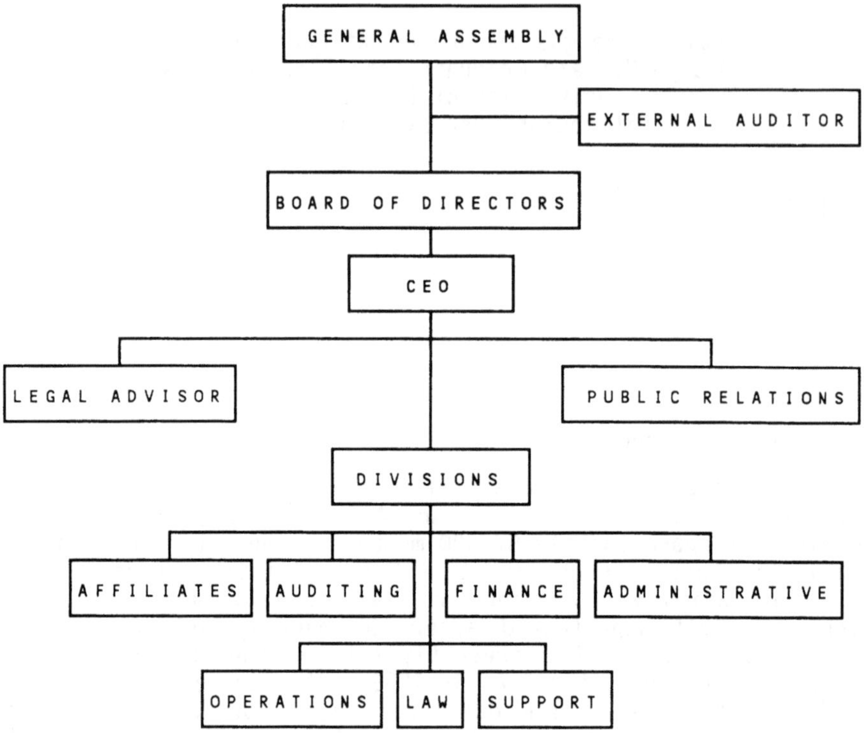

elsewhere. However, the sectors of the population that each department serves are different. For example, Normal Operations distributes funds to qualified manufacturing and industrial companies. It also provides financing to help individuals maintain property, ease cash flow problems, and satisfy primary home needs. Popular Credit allocates funds to small businesses that provide basic-need products such as food and clothing. The other two funds provide loans to employees in the public or private sector (the EMS fund) and to employees of FDC (the PTN fund). The PTN fund is also used to provide benefits for retired employees and their families.

In 1985, financial support from other institutions that supplied funds to FDC decreased dramatically. The purpose of this project, carried out in 1986, was to help FDC's management to develop a long-term strategic plan (1991) and short-term (1987) objectives to regain FDC's credibility and improve its financial efficiency. To develop a strategy, a group of selected individuals analyzed the organization's operations and the scenarios which FDC was likely to encounter between 1986 and 1991.

2. ANALYSIS OF FDC'S CURRENT SITUATION

Until 1986, the strategy of FDC had been one of expansion: FDC increased the dollar amount of loans and created new SLAs to cover as much of the country's territory as possible. The implementation of this strategy was a consequence of the political interests of the President of the Republic and the CEO. By expanding, FDC showed the commitment of the government to support small businesses such as farmers. Support of these groups, which comprise more than 80% of the country's population, translates into favorable votes for the government.

The expansionary strategy of FDC was at best questionable given its financial situation at that time. Table 1 summarizes the Income Statement for the year ended December 31, 1985. Table 2 gives the Balance Sheet as of December 31, 1985. Table 3 shows how all the resources were distributed among the various funds, and Table 4 shows the changes in the financial position of FDC from 1984 to 1985. While the total amount of funds available decreased by 33 percent, the total amount of funds allocated to loans increased by 27 percent. These changes produced a net decrease in FDC's working capital of 48 percent.

A comparison of FDC's financial ratios with the same ratios for the average U.S. Savings and Loan Association of equal size is illustrated in Table 5. The data given in this table was published in 1987 but pertains to 1986. There is no significant difference between these ratios and the 1985 ratios. Although economic requirements and government regulations for financial institutions vary between the two countries, the differences between some of the ratios serve to illustrate the financial position of FDC. Indeed, the most striking difference is shown by the total liability to net worth ratio. Due to the high percentage of delinquent loans, FDC needs a much larger equity base relative to its debt. FDC's debt ratio is only 77.23 percent compared to 96.8 percent for Savings and Loans Associations in the U.S. Its ROA and ROE are 0.3% and 1.3%, respectively, which compare unfavorably with the U.S. industry ratios of 0.5% and 9.8%, respectively. FDC's low ROA indicates an inefficient utilization of their assets to generate profits.

In addition, their ability to pay long-term debt was very limited as reflected by the coverage ratio of 1.06 (i.e., the net income only covers 6 percent of the interest that must be paid on borrowed funds). FDC's questionable ability to pay debt combined with a high percentage of delinquent loans were the most likely reasons for the reduction of loans by the Republic's Federal Reserve Bank to FDC (from $34.28 million in 1984 to $8.34 million in 1985). In summary, FDC's substantially lower leverage was due to the organization needing much more equity to compensate for fraudulent loans and to sustain a desired degree of liquidity. Expansionary policies could only be sustained from within the organization by increasing the equity base or by reducing the risk of unpaid loans. Both measures would allow FDC to increase the total dollar amount of loans and the number of approved loans.

Table 1. Income Statement For the Year Ended December 31, 1985 ($USA millions)

	FDN Fund		EMS Fund		PTN Fund		TOTAL	
REVENUES								
- Financial Operation								
Interest on Loans	3.29		0.86		0.20		4.35	
Interest on Deposits	0.39						0.39	
Fees and Commissions	0.001	3.681		0.86		0.20	0.001	4.741
- Administrative and Services		0.19		0.04				0.23
- Non-Financial Operations								
Sale of Fertilizers	4.21						4.21	
Other Sales	0.29						0.29	
Other	0.031	4.531					0.031	4.531
TOTAL REVENUES		**8.402**		**0.90**		**0.20**		**9.502**
EXPENSES								
- Financial								
Interest in Debts	2.46		0.42				2.88	
Commissions and Fees	0.07	2.53	0.01	0.43			0.08	2.96
- Administrative Expenses		1.90		0.45		0.09		2.44
- Non-Financial Expenses								
Sale of Fertilizers	3.67						3.67	
Other	0.27	3.94					0.27	3.94
TOTAL EXPENSES		**8.37**		**0.88**		**0.09**		**9.34**
NET INCOME		**0.032**		**0.02**		**0.11**		**0.162**

Table 2. Balance Sheet As of December 31, 1985 ($USA millions)

	FDN Fund		EMS Fund		PTN Fund		TOTAL	
ASSETS								
- Cash and Due from Banks		1.39		0.41		0.68		2.48
- Investment Securities		1.01						1.01
- Loans								
Short-Term	9.35		0.01		0.31		9.67	
Medium-Term	14.42		8.19		0.37		22.98	
Long-Term	7.48				0.80		8.29	
Less Allowance for possible credit losses	(0.30)	30.95		8.20		1.48	(0.30)	40.63
- Premises and Equipment		2.69	0.02					2.71
- Other Assets		7.49	0.01			0.04		7.54
TOTAL ASSETS		**43.53**		**8.64**		**2.20**		**54.37**
LIABILITIES								
- Deposits	2.02						2.02	
- Acceptances Outstanding	1.23		0.46				1.96	
- Circulating Bonds	4.44		1.43				5.37	
- Federal Reserve Funds Borrowed	19.94						19.94	
- Other Local Funds Borrowed	3.43		3.63				7.06	
- Foreign Funds Borrowed	2.64						2.64	
- Other Liabilities	2.20		0.55		0.02		2.77	
TOTAL LIABILITIES		35.90		6.07		0.02		41.99
STOCKHOLDERS' EQUITY								
- Common Stock & Paid-in-Capital	6.00		0.21		1.57		7.78	
- Other Paid-in-Capital (Reserves)	1.20		1.85				3.05	
- Retained Earnings	0.43		0.51		0.61		1.55	
TOTAL STOCKHOLDERS' EQUITY		7.63		2.57		2.18		12.38
TOTAL LIABILITIES & STOCKHOLDERS' EQUITY		**43.53**		**8.64**		**2.20**		**54.37**

Table 3. Allocation of Funds by Project Categories

	% of Total	
FDN Fund	77.88	
- Agriculture		16.38
- Normal Operations		4.96
- Urban Development		5.40
- Rural Development		22.88
- Popular Credit		28.26
EMS Fund	19.11	
PTN Fund	3.01	
TOTAL	100.00	

Table 4. FDC's Statement of Changes in Financial Position Working Capital Basis For the Year Ended December 31, 1985 ($USA millions)

SOURCES OF FUNDS	85	84	VARIATION
Available Initial Funds	4.70	7.26	(2.56)
Paid Back Loans	22.30	18.92	3.38
Funds from the Federal Reserve Bank	8.34	34.28	(25.94)
Other Sources Funds	2.44	3.22	(.78)
Special Operations Funds	4.52	2.42	2.10
Deposits and Financial Operations	4.94	4.76	0.18
	47.24	70.86	(23.62)
USES OF FUNDS			
Loans	23.18	18.22	4.96
Investments	0.32	0.36	(.04)
Amortization	9.92	37.54	(27.62)
Special Operations	3.94	2.0	1.94
Purchase of Equipment	.06	.04	.02
Construction	.12	-----	.12
Administrative Expenses	2.44	1.92	.52
Financial Expenses	2.98	2.7	.28
Final Availability of Funds	2.48	4.68	(2.2)
	45.44	67.46	(22.02)
Working Capital	**1.78**	**3.4**	**(1.62)**

Table 5. Firms With 50M to 100M in Assets

	FDC**	SAVINGS & LOANS* ASSOCIATIONS IN THE USA (Average)
1. **Debt Ratio** (Total Debt/Total Assets)	77.23	96.8%
2. **ROA** (Net Income/Total Assets)	0.3%	0.5%
3. **ROE** (Net Income/Stockholders' Equity)	1.3%	9.8%
4. **Coverage Ratio** ((Net Income + Interest Paid)/Interest Paid)	1.06 to 1	1.1 to 1
5. **Total Liability to Net Worth** (Total Liabilities/Stockholders' Equity)	3.4 to 1	30.2 to 1

* Source: Troy, L., Almanac of Business and Industrial Ratios, 1987.
** For December 31, 1985.

Table 6. Total and Delinquent Loans For the Year Ended December 31, 1985 ($USA millions)

FUNDS	Total Loans 1984		Total Loans 1985		Delinquent Loans		% of Total	
FDN Fund								
- Agriculture	2.82		3.80		5.54		47.11	
- Normal Operations	2.41		1.15		2.60		22.11	
- Urban Development	0.67		1.25		0.56		4.74	
- Rural Development	4.47		5.30		1.84		15.65	
- Popular Credit	4.71		6.55		0.07		0.60	
		15.08		18.05		10.61		**90.22**
EMS Fund		2.75		4.42		1.09		**9.27**
PTN Fund		0.42		0.71		0.06		**0.51**
TOTAL		**18.25**		**23.18**		**11.76**		100.00

The single major problem of FDC was the high delinquency rate of its borrowers. At the end of 1985 there were nearly $12 million in accumulated delinquent loans (see Table 6).

In 1985 the delinquency rate was above 10 percent of the total amount granted in loans. Forty seven percent of the delinquent loans came from the agricultural programs in the FDN Fund. The main cause of the high delinquency problem was the unstable political situation caused by the existing war. The political instability and the internal civil struggle between opposing factions together with the high unemployment and the new institutionalized agricultural reform created a turbulent environment in which financial institutions struggled to survive. The 1980 agricultural reform eliminated landlords with large land holdings and converted hourly workers into land owners. FDC organized seminars to educate the new land owners on financial responsibilities and land administration.

Despite measures taken by FDC's management to curtail delinquency and recuperate bad loans (e.g., more stringent criteria to approve loans, increasing the loan collection work force, etc.) FDC experienced an "image problem." The Federal Reserve Bank of the Republic would have been more willing to provide additional funds at a lower rate if the delinquency rate had dropped and if the SLAs had functioned more like true banking institutions with checking accounts, international exchange, etc. rather than as distributors of funds. In addition, international agencies such as AID stopped the flow of funds because FDC had adopted an expansionary strategy that could not be supported by its financial position.

3. DEVELOPMENT OF A BUSINESS STRATEGY FOR FDC

Despite FDC's difficult situation and its need for more funds, the CEO was reluctant to change the organization's current strategy of creating new SLAs and providing more loans. However, to qualify for additional funds from an international agency, the Board of Directors agreed to develop a strategy that would address some of FDC's crucial problems. To develop a new strategy, a decision-making group, consisting of members of the General Assembly, divisional directors, and several lower level managers, was convened. This group exhibited a wide variety of decision-making skills drawn from the business world and engineering discipline. The group, along with several consultants, required two decision-making sessions of one and one half days duration to model FDC's problem.

The Analytic Hierarchy Process (AHP) was chosen as the most appropriate tool to model the new strategic plan for the following reasons:

- the AHP is systematic;
- it encourages creativity;
- qualitative factors can be handled;
- group decision-making is facilitated; and
- the process is flexible and allows for updates.

The first step of the decision-making process was to create scenarios (Status Quo, Pessimistic, and Optimistic) for both FDC and the entire country. The Status Quo scenario represents the present state of the environment in which FDC operates. The other two scenarios reflect extreme deviations from the present state. Each scenario consists of three dimensions, Economic, Financial, and Sociopolitical, and each dimension is further broken down. These scenarios reflect the decision-making group's perception of the economic, financial, and sociopolitical situations of both the country and FDC. For example, in the economic dimension, the members of the group thought that resources from the sale of fertilizers would increase in the Optimistic scenario and remain the same or decrease in the Status Quo or Pessimistic scenarios. Since FDC depends on 80 percent of its resources from external sources, they predicted that the best that could happen to FDC was the continuation of the same level of dependence on external sources of funds. Factors referring to inflation and bad debt were assigned numerical values. However, the majority of the factors were qualitatively described.

The next step in this process was to identify FDC's global strategies. One way of doing this is to match the weaknesses and strengths of the agency with the resources that can be used to attain the goals. These resources may be obtained from the organization itself or from external sources.

Because FDC is a governmental organization and its image has been tarnished by the large number of delinquent loans, strategies involving acquisitions or mergers with other more solvent financial institutions were not realistic. The group identified the following weaknesses and strengths of FDC:

STRENGTHS	WEAKNESSES
- Existing Network	- Organizational Structure
- Executive and Employee Programs	- Communications System
- Image with the Consumer	- Data Processing
- Fertilizers	- Salaries

Being aware of the FDC's strengths and weaknesses prompted the group to formulate three long-term (1991) objectives. They were to:

-O_1 Obtain an adequate lending rate from the Federal Reserve Bank of the Republic,

-O_2 Decrease the delinquency rate to less than 3% (i.e., $0.762 million in 1985), and

-O_3 Decrease administrative costs.

It became clear to the participants that the strategies they should consider had to address the three issues stated in the long-term objectives. Matching the weaknesses and strengths with the long-term objectives, the group arrived at the following global strategies:

-ST1 Improve the organization's image,

-ST2 Obtain negotiation capabilities, i.e., the ability to obtain funds to support FDC's programs,

-ST3 Ensure administrative autonomy for the SLAs,

-ST4 Continue the current strategy with a business emphasis, and

-ST5 Pursue an educational emphasis.

The main objective of the first two strategies was to obtain an adequate lending rate from the Federal Reserve Bank. The remaining three strategies address the other two objectives. Thus, ST3 would allow individual SLAs to be true banking institutions and pursue some of the objectives of their managers. ST4 would provide similar capabilities as ST3 except that the central administration of FDC would still make most of the crucial decisions. Finally, ST5's major objective is to educate the population. A long-term result of ST5 would be the decline of the delinquency rate because the borrowers would have better opportunities (e.g., starting new joint ventures, developing new products and services, securing funding from other sources) after obtaining further education on business administration, farming processes, and industrial product commercialization.

All of the objectives were important to FDC and no single strategy addressed all of the long-term objectives at the same time. As a result, FDC's strategy should be a mix of the five basic strategies. In a single objective environment only one strategy would have been chosen. In the multi-objective situation that faced the decision-making group, the "best" strategy had to be shaped by making tradeoffs among the long-term objectives and by taking into account the likelihood of occurrence of the scenarios. To determine this combination, the group ranked the long-term objectives in terms of the scenarios and the strategies in terms of the long-term objectives. The structure used to represent this part of the process is shown in Figure 2.

Next, the long-term objectives were ranked with respect to the scenarios. The pairwise comparisons are shown in Figure 3.

The two matrices and the priorities that result from them indicate that if the future of FDC and the country is either a CONTINUATION of the present or it DETERIORATES further, then the long-term objective that is most important is to DECREASE THE DELINQUENCY RATE (O_2). However, if OPTIMISTIC is the most likely scenario, then the most important objective is to OBTAIN AN ADEQUATE LENDING RATE (O_1).

Figure 2. A Hierarchy to Develop Strategy

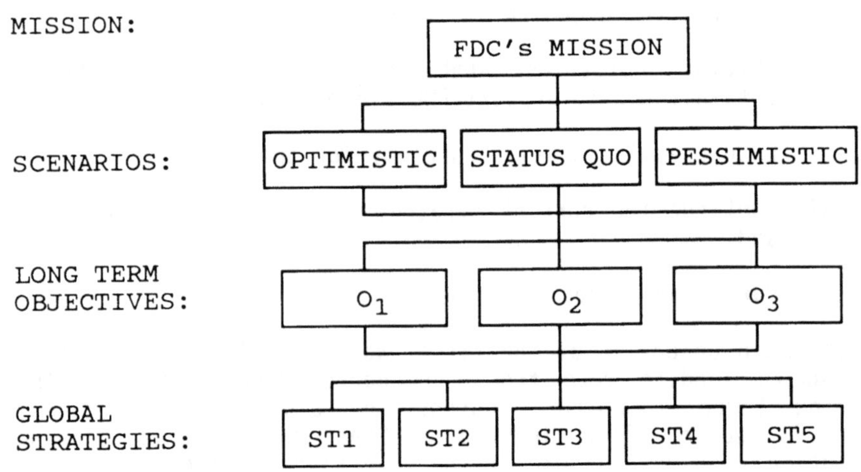

It was the opinion of the group that the most likely scenario would be the STATUS QUO, although the OPTIMISTIC scenario was not completely unlikely. Thus, they did not rank the scenarios according to the likelihood of occurrence but assumed that the Status Quo scenario was the most likely. However, to develop a strategy, all of the scenarios were considered.

Next, the group ranked the strategies identified according to the long-term objectives for each of the scenarios. To accomplish this, the participants agreed that it would not be worthwhile to pursue some of the strategies under some of the scenarios. For example, under the Status Quo and Pessimistic scenarios, strategies ST1, ST2, and ST3 were equally important but irrelevant in comparison to strategies ST4 and ST5. The judgments used to obtain the priorities of ST3, ST4, and ST5 according to O_2 and O_3 are given in Figure 4.

Figure 3. Pairwise Comparisons and Priorities of Long-Term Objectives According to the Scenarios

SCENARIOS (ST.QUO, PESSIMISTIC)	O_1	O_2	O_3	PRIORITIES
O_1	1	*	*	0.08
O_2	5	1	5	0.69
O_3	5	*	1	0.23

OPTIMISTIC SCENARIO	O_1	O_2	O_3	PRIORITIES
O_1	1	5	5	0.70
O_2	*	1	3	0.10
O_3	*	*	1	0.20

* = reciprocal of the number in the transposed position.

Figure 4. Priorities of Strategies According to the Three Scenarios and the Long-term Objectives O_2 and O_3

(STATUS QUO, PESSIMISTIC, OPTIMISTIC)

O_2, O_3	ST3	ST4	ST5	PRIORITIES ST.Q.	PESS.	OPT.
ST3	1	*	*	0.08	0.33	0.05
ST4	(7,1,9)	1	*	0.49	0.33	0.50
ST5	(5,1,7)	(1,1,1)	1	0.43	0.33	0.45

From Figure 4, we see that these three strategies have almost the same priorities under the Status Quo scenario as under the Optimistic scenario, and that the fourth and fifth strategies are more important than the third one. The three strategies are equally important in the Pessimistic scenario.

The global or composite priorities of the strategies are shown in Figure 5.

Figure 5. Composite Priorities of the Strategies

STRATEGIES	LONG-TERM OBJECTIVES O_1	O_2	O_3			PESSIMISTIC		GLOBAL PRIORITIES
ST1	.50	-	-					0.04
ST2	.50	-	-		O_1	0.08		0.04
ST3	.00	.33	.33	x	O_2	0.69	=	0.31
ST4	.00	.33	.33		O_3	0.23		0.31
ST5	.00	.33	.33					0.31

STRATEGIES	LONG-TERM OBJECTIVES O_1	O_2	O_3			STATUS QUO		GLOBAL PRIORITIES
ST1	.50	-	-					0.04
ST2	.50	-	-		O_1	0.08		0.04
ST3	.00	.08	.08	x	O_2	0.69	=	0.07
ST4	.00	.49	.49		O_3	0.23		0.45
ST5	.00	.43	.43					0.40

STRATEGIES	LONG-TERM OBJECTIVES O_1	O_2	O_3			OPTIMISTIC		GLOBAL PRIORITIES
ST1	.50	-	-					0.35
ST2	.50	-	-		O_1	0.70		0.35
ST3	.00	.05	.05	x	O_2	0.10	=	0.02
ST4	.00	.50	.50		O_3	0.20		0.15
ST5	.00	.45	.45					0.14

These priorities suggest the following retrenchment strategy:

If the political struggle continues and the operating environment is a prolongation of the present (Status Quo scenario) or worsens (Pessimistic scenario), then FDC should shift its strategy from opening more SLAs without sufficient funds to transforming the current SLAs into more business-oriented subsidiaries without opening new ones and continue with the educational programs in which it was already involved. On the other hand, if the operating environment improves, then they should attempt to improve the image of the organization among the banking institutions and to acquire some skills in negotiation to deal with the Federal Reserve Bank. At this stage, FDC should not attempt to expand beyond present levels, but should try to reenforce its present capabilities.

4. IMPLEMENTATION OF THE STRATEGY

The most important step in strategic planning is the transformation of strategic thinking into strategic action. This transformation takes place by

- identifying measurable annual objectives,
- developing functional strategies, and
- formulating simple policies to guide the development.

To accomplish this transformation, the group identified some short-term (1987) objectives for each of the long-term objectives mentioned previously. To help attain these short-term objectives, the group also developed some functional strategies. Both short-term objectives and functional strategies are listed in Table 7. For example, the long-term objective of obtaining an adequate lending rate (O_1) could only be negotiated in the Optimistic scenario which was considered unlikely to occur and, hence, no short-term objectives were identified. However, the group identified two functional strategies that would help attain the objective once the environment changed from the Status Quo to the Optimistic scenario. The first strategy of designing a negotiation policy refers to FDC's need of assembling a team of skilled people to negotiate with the Federal Reserve Bank and other institutions. The second strategy develops a document that states the image problem and identifies specific actions that could help improve it.

To address the problem of delinquent loans, the group formulated the short-term objective of obtaining ninety percent of the outstanding debt within thirty days and improving the collection of loans by 15 percent in 1987. During the process of identifying short-term objectives for 1987, they realized that over 95 percent of loan applications were for amounts of $2,000 or less. It had been the policy of FDC to approve all loans that the SLAs recommended. The group decided to implement a strategy in which the SLAs would give final approval for all loans under $2,000.

Table 7. Short-Term Objectives (1987) and Functional Strategies

Long-Term Objective	Obtain an adequate spread
Functional Strategies	Design a negotiation policy
	Develop a document addressing the problem
Long-Term Objective	Develop a process to decrease the loss due to bad debt
Short-Term Objectives	Obtain 90% of the outstanding debt within the period of 30 days available to pay the fees of the new loans
	Improve the collection of loans by 15% in 1987
Functional Strategy	The SLAs will approve all loans for amounts less than $2,000 without consultation with the central administration
Long-Term Objective	Minimize administrative costs
Short-Term Objectives	Decrease costs per SLA by 0.5% in 1987
	Increase the efficiency of the system without laying off personnel
Functional Strategy	Keep the same level of employment within the organization

Finally, the group formulated short-term objectives to increase the efficiency of FDC by decreasing administrative costs. However, due to the high unemployment in the nation it was decided that the level of employment would be maintained within FDC.

5. CONCLUSIONS

The principal outcome of the project was a document that the group of participants presented to the CEO of the organization. The document suggested a change in FDC's strategy. The new strategy recommended that FDC should stop expanding and that the existing SLAs should be allowed the freedom to act as banking institutions rather than as distributors of funds. In addition, as a result of the strategy-making sessions, new funds (approximately $5 million) were allocated to FDC from an international agency.

Although the experience was very valuable to both the authors and the participants, we failed to foresee that the decision-making group did not have sufficient power to fully implement the recommended strategy. The CEO declined to change FDC's strategy because of political considerations. Nonetheless, the group benefited from the AHP-based process and found it enlightening and very helpful in focusing the discussion. It was the first time that all levels of management had worked together to address FDC's problems.

MIX
Papier aus verantwortungsvollen Quellen
Paper from responsible sources
FSC® C105338
FSC
www.fsc.org

If you have any concerns about our products,
you can contact us on
ProductSafety@springernature.com

In case Publisher is established outside the EU,
the EU authorized representative is:
Springer Nature Customer Service Center GmbH
Europaplatz 3, 69115 Heidelberg, Germany

Printed by Libri Plureos GmbH
in Hamburg, Germany